Siddhartha Bhattacharyya, Hrishikesh Bhaumik, Anirban Mukherjee,
Sourav De (Eds.)

Machine Learning for Big Data Analysis

De Gruyter Frontiers in Computational Intelligence

Edited by Siddhartha Bhattacharyya

Volume 1

Machine Learning for Big Data Analysis

Edited by Siddhartha Bhattacharyya,
Hrishikesh Bhaumik, Anirban Mukherjee, Sourav De

DE GRUYTER

Editors

Prof. (Dr.) Siddhartha Bhattacharyya
RCC Institute of Information Technology,
Canal South Road, Beliaghata,
Kolkata 700 015, India
dr.siddhartha.bhattacharyya@gmail.com

Mr. Hrishikesh Bhaumik,
RCC Institute of Information Technology
Canal South Road, Beliaghata,
Kolkata 700 015, India
hbhaumik@gmail.com

Dr. Anirban Mukherjee
RCC Institute of Information Technology
Canal South Road, Beliaghata
Kolkata 700 015, India
anirbanm.rcciit@gmail.com

Dr. Sourav De
Cooch Behar Government Engineering College
Vill- Harinchawra, P.O.- Ghughumari, Cooch
Behar - 736170
West Bengal India
sourav.de79@gmail.com

ISBN 978-3-11-055032-0
e-ISBN (PDF) 978-3-11-055143-3
e-ISBN (EPUB) 978-3-11-055077-1
ISSN 2512-8868

Library of Congress Cataloging-in-Publication Data
Names: Bhattacharyya, Siddhartha, 1975- editor. | Bhaumik, Hrishikesh, 1974- editor. | Mukherjee, Anirban, 1972- editor. | De, Sourav, 1979- editor.
Title: Machine learning for big data analyis / edited by / Herausgegeben von Siddhartha Bhattacharyya, Hrishikesh Bhaumik, Anirban Mukherjee, Sourav De.
Description: Berlin : Walter de Gruyter GmbH, [2018] | Series: Frontiers in computational intelligence ; volume 1 | Includes bibliographical references and index.
Identifiers: LCCN 2018031292 (print) | LCCN 2018033492 (ebook) | ISBN 9783110551433 (electronic Portable Document Format (pdf) | ISBN 9783110550320 (print : alk. paper) | ISBN 9783110551433 (ebook pdf) | ISBN 9783110550771 (ebook epub)
Subjects: LCSH: Big data. | Machine learning. | Quantitative research.
Classification: LCC QA76.9.B45 (ebook) | LCC QA76.9.B45 M33 2018 (print) | DDC 005.7–dc23
LC record available at https://lccn.loc.gov/2018031292

Bibliographic information published by the Deutsche Nationalbibliothek
The Deutsche Nationalbibliothek lists this publication in the Deutsche Nationalbibliografie; detailed bibliographic data are available on the Internet at http://dnb.dnb.de.

Cover image: shulz/E+/getty images
Typesetting: le-tex publishing services GmbH, Leipzig
Printing and binding: CPI books GmbH, Leck

www.degruyter.com

Preface

Big data is a term used to describe data sets that are so large or complex that traditional data processing application software is inadequate to deal with them. The possible challenges in this direction include capture, storage, analysis, data curation, search, sharing, transfer, visualization, querying, updating and information privacy. The term "big data" often refers simply to the use of predictive analytics, user behaviour analytics, or certain other advanced data analytics methods that extract meaningful value from data without concern for the size of the data set. Due to the advances in data evolution, scientists are encountering limitations in e-Science work, including meteorology, genomics, connectomics, complex physics simulations, biology and environmental research.

Big data analytics is the process of examining large and varied data sets – i.e., big data – to uncover hidden patterns, unknown correlations, market trends, customer preferences and other useful information that can help organizations make more informed business decisions. Big data analytics applications enable data scientists, predictive modellers, statisticians and other analytics professionals to analyse growing volumes of structured transaction data, plus other forms of data that are often left untapped by conventional business intelligence (BI) and analytics programs that encompasses a mix of semi-structured and unstructured data. On a broad scale, data analytics technologies and techniques provide a means of analysing data sets and drawing conclusions about them to help organizations make informed business decisions. BI queries answer basic questions about business operations and performance. Big data analytics is a form of advanced analytics that involves complex applications with elements such as predictive models assisted by statistical algorithms powered by high-performance analytics systems.

Note that unstructured and semi-structured data of these types typically do not fit well in traditional data warehouses that are based on relational databases oriented towards structured data sets. Furthermore, data warehouses may not be able to handle the processing demands posed by sets of big data that need to be updated frequently – or even continually, as in the case of real-time data on stock trading, the online activities of website visitors or the performance of mobile applications. As a result, many organizations that collect, process and analyse big data turn to Hadoop and its companion tools, such as YARN, MapReduce, Spark, HBase, Hive, Kafka and Pig, as well as NoSQL databases. In some cases, Hadoop clusters and NoSQL systems are used primarily as landing pads and staging areas for data before they get loaded into a data warehouse or analytical database for analysis, usually in a summarized form that is more conducive to relational structures. Of late, scientists and researchers have resorted to machine intelligence for analysing big data, thereby evolving BI. It is a well-known fact that data in any form exhibit varying amounts of ambiguity and impreci-

https://doi.org/10.1515/9783110551433-201

sion. Machine learning tools and strategies are adept at handling these uncertainties and, hence, extracting relevant and meaningful information from data.

This volume comprises six well-versed contributed chapters devoted to reporting the latest findings on the applications of machine learning for big data analytics.

Chapter 1 provides a hands-on introduction to psychometric trait analysis and presents a scalable infrastructure solution as a proof of concept for two important concepts, efficient handling of enormous amounts of available data and the demand for micro-targeting. The authors discuss two use cases and show how psychometric information, which could, for example, be used for targeted political messages, can be derived from Facebook data. Finally, potential further developments are outlined that could serve as starting points for future research.

Video summarization is an important field of research in content-based video retrieval. One of the major aims in this domain has been to generate summaries of videos in the shortest possible time. In Chapter 2, the primary aim is to rapidly select keyframes from the composing shots of a video to generate a storyboard in a minimal amount of time. The time taken to produce the storyboard is directly proportional to the number of correlations to be computed between the frames of a shot. To reduce the time, user input is obtained regarding the amount of actual correlations to be computed. Keyframes are selected from each shot by generating an approximate minimal spanning tree and computing the density around each frame of the shot by means of an automatic threshold based on the statistical distribution of the correlation values.

Most techniques for image processing involves algorithms that are custom built and lack flexibility, making them different to the data being processed. In Chapter 3, the authors elaborate upon various methodologies within the domain of image processing. They chronologically demonstrate the role of learning techniques involved in image super resolution, image upsampling, image quality assessment and parallel computing techniques. Further, an in-depth explanation is provided of the involvement of deep neural architectures as an impressive tool for performing multiple image processing tasks.

Chapter 4 focuses on connected cities in terms of smart transportation. A connected city controls available resources in such a way that it can efficiently improve economic and societal outcomes. Many data are generated from people, systems and things in cities. Thus, data generated from various resources are considered to be the most scalable asset of a connected city. Heterogeneous data are difficult to organize, interpret and analyse. Generally, the data generated from various sources are very large and heterogeneous as well because of they are generated from heterogeneous environments like water, traffic, energy, buildings and so forth. Hence, different multidimensional contexts like databases, data mining, artificial intelligence and distributed systems communities are useful for dealing with the challenges of big data in connected cities.

A new hybrid structure of neural fuzzy networks is proposed and studied in Chapter 5, combining through the layer of fuzzy clustering a fuzzy cellular Kohonen neural network and radial-basic neural network. The proposed model has a high degree of self-organization of neurons, improving the separation of network properties in case of overlapping clusters; automatic adjustment of the parameters of radially symmetric functions; the presence of a single hidden layer, sufficient for modelling pronounced non-linear dependencies; a simple algorithm for optimizing weight coefficients; and a high learning speed. The model can be used to solve a wide range of problems – clusterization, approximation and classification (recognition) of multidimensional, semi-structured data.

Image fusion is a combination of multiple images that results in a fused image. It provides more information than any other input images and is based on discrete wavelet transformation (DWT) and sparse sampling. The sparse sampling method offers better performance than the Nyquist theorem for signal processing. Among the various techniques, DWT offers many advantages; it yields higher quality and requires less storage and low cost, which is very useful in image applications. Image-related applications have few constraints such as minimum data storage and less bandwidth for communication that takes place through satellite, which actually results in the capture of low-quality images. To overcome this problem, image fusion has proven to be a potential tool for remote sensing applications that incorporate data from combinations of panchromatic, multispectral images; for bringing out a composite image having both higher spatial and spectral resolutions. The research in this area goes back a couple of decades; the diverse approaches and methodologies proposed so far by the various researchers in the field are discussed in Chapter 6.

This volume is intended to be used as a reference by undergraduate and postgraduate students in the fields of computer science, electronics and telecommunication, information science and electrical engineering as part of their curriculum.

April 2018
Kolkata, India

Siddhartha Bhattacharyya
Hrishikesh Bhaumik
Anirban Mukherjee
Sourav De

Siddhartha Bhattacharyya would like to dedicate this book to his late father Ajit Kumar Bhattacharyya, his late mother Hashi Bhattacharyya, his beloved wife Rashni Bhattacharyya and his cousin sisters-in-law Aparna Bhattacharjee and Baby Bhattacharjee

Hrishikesh Bhaumik would like to dedicate this book to his late father Major Ranjit Kumar Bhaumik, who is his greatest inspiration, and to his mother Mrs Anjali Bhaumik, who has supported and stood by him through all the ups and downs of life

Anirban Mukherjee would like to dedicate this book to respected Professor Sujit Ghosh

Sourav De would like to dedicate this book to his son Aishik, his wife Debolina Ghosh, his father Satya Narayan De, his mother Tapasi De and his sister Soumi De

Contents

Till Blesik, Matthias Murawski, Murat Vurucu, and Markus Bick

1 Applying big data analytics to psychometric micro-targeting

Abstract: In this chapter we link two recent phenomena. First, innovations in technology have lowered the cost of data storage and enabled scalable parallel computing. Connected with social media, the Internet of Things applications and other sources, large data sets can easily be collected. These data sets are the basis for greatly improving our understanding of individuals and group dynamics. Second, events such as the election of Donald J. Trump as President of the United States of America and the exit of Great Britain from the European Union have shaped public debates on the influence of psychometric micro-targeting of voters. Generally, public authorities, but also other organizations, have a very high demand for information about individuals.
We combine these two streams, meaning the enormous amounts of data available and the demand for micro-targeting, aiming at answering the following question: *How can big data analytics be used for psychometric profiling?* We develop a conceptual framework of how Facebook data might be used to derive the psychometric traits of an individual user. Our conceptual framework includes the Facebook Graph API, a nonSQL Mongo Data Base for information storage and R scripts to reduce the dimensionality of large data sets by applying the latent Dirichlet allocation to determine correlations between reduced information with psychologically relevant words.
In this chapter we provide a hands-on introduction to psychometric trait analysis and present a scalable infrastructure solution as a proof of concept for the concepts presented here. We discuss two use cases and show how psychometric information, which could, for example, be used for targeted political messages, can be derived from Facebook data. Finally, potential further developments are outlined that could serve as starting points for future research.

Keywords: Big data, Big Five personality traits, Facebook, Politics, Psychometrics, Latent Dirichlet allocation

1.1 Introduction

Technological innovations in the twentieth and twenty-first centuries have had immense impacts on society. The emergence of the Internet and the resulting permanent connectivity of individuals has changed not only the economy but also the way society functions in general [1]. Before the infiltration of measurable online actions, data were

Till Blesik, Matthias Murawski, Murat Vurucu, Markus Bick, ESCP Europe Business School, Heubnerweg 8–10, 14059 Berlin, Germany, e-mails: {tblesik, mmurawski, mbick}@escpeurope.eu, murat.vurucu@edu.escpeurope.eu

https://doi.org/10.1515/9783110551433-001

scarce. That is why statistical inference, analysing a data set derived from a larger population, was and still is very important. It helps make the best of scarce and expensive data.

Now, in the age of social media, the Internet of Things, e-commerce, online financial services, search engines, navigation systems and cloud computing, data are being collected from every individual to machines and processed by machine-to-machine interactions [2]. These data can be analysed in terms of joint correlations, which generates even more data, called meta-data. A single smartphone alone is already able to provide information about the purchasing habits, transportation preferences and routes, personal preferences and social surroundings of individual users.

It is not far-fetched to imagine how public authorities use their applications to listen to spoken words and translate them to written and, therefore, searchable text, mapping all movements and patterns of an individual and using correlations of in-between behaviour and preferences collected from social networks to identify, for instance, potential threats to the state. All of the international invasions of privacy carried out by the US National Security Agency (NSA) and similar organizations have shaped the public debate and our perception of technology dramatically since the Edward Snowden leaks in 2013. An Orwellian fantasy of mass surveillance seems to have become a reality in the shape of modern government.

Secret services are, however, also arguably relevant institutions within democracies. Based on a judicial system that follows a democratic constitution and aims at protecting a nation's interests, secret services build their operations on judges and laws and usually use their applications in a political context of checks and balances to categorize and control enemies of the state. The nation's "interest" and the nation's "enemies" are, however, terms whose definitions greatly depend on the ideas of the current elected government. In fact, given the technology already in place, today's governments are potentially able to build "psychometric" profiles of every single voter so as to influence their voting habits [3].

Psychometrics is a scientific approach to assessing the psychological traits of people. It has its roots in sociobiology [4]. The goal is to obtain a distribution within the population of each of the personality traits of the Big Five personality test. The Big Five traits are *extraversion*, *agreeableness*, *openness to experience*, *conscientiousness* and *emotional stability/neuroticism* [5, 6]. There are various ways to explore individual traits, for example by analysing the Facebook likes of a user and other forms of written texts such as status messages. This information can be connected to other demographic data such as age or gender. In the context of politics, messages sent to voters can potentially be adapted, based on the results of a psychometric analysis. A governor advocating lax gun laws would address a young mother in a different way than a gun enthusiast in the National Rifle Association (NRA). The young mother might receive a message advocating lax gun laws so that teachers can carry guns in educational settings to protect her children, while an NRA member would receive a message demonstrating the newest features of a military weapon that should be legalized.

Understanding the psychometric traits of each recipient can help define the content of a message. Psychometric analysis has been used for over a century, but in the context of big data and mass surveillance, it has gained new importance [4].

Based on these thoughts, we will investigate the opportunities for deriving the psychometric traits of individual users from Facebook data. More precisely, the research question can be summarized as follows:

How can Facebook data be extracted, stored, analysed, presented and used for micro-targeting within the Big Five model?

To answer this question, several data sources and calculations are used, as shown in Figure 1.1.

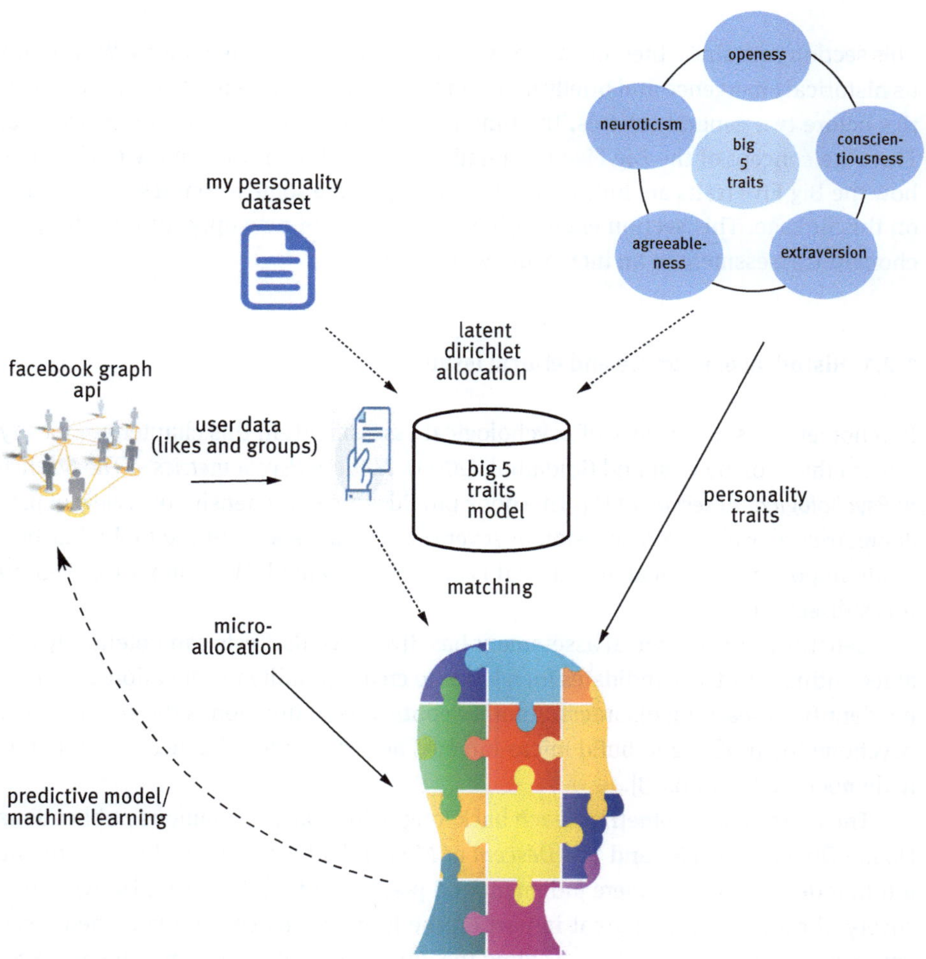

Fig. 1.1: Conceptual flowchart of this chapter

This chapter is structured as follows. The second section introduces the theoretical and historical foundations of psychometric analysis. The third section presents the research methodology while placing a specific focus on both the underlying statistical and technical infrastructure, including a presentation of how Facebook can be used as a data source for our study. We test our conceptual framework in Section 1.4, which includes two use cases covering the preparation of data, the extraction of patterns and corresponding final results. This chapter ends with a discussion of our results, the limitations of our approach and some concluding remarks.

1.2 Psychometrics

This section contains a theoretical overview of the topic of psychometrics. We present its historical emergence and briefly mention some ethical issues related to psychometrics before two general schools, the functional and the trait schools, are discussed. Then the concept of the Big Five personality traits will be introduced. We will show how the Big Five traits are linked to politics and provide some recent research results on this linkage. This section ends with a presentation of new opportunities for psychometric assessment in an increasingly digital world.

1.2.1 Historical emergence and ethical issues

Psychometrics is the science of psychological assessment. In this chapter, we mostly refer to the book by Rust and Golombok (2009), *Modern Psychometrics – The Science of Psychological Assessment* [4]. This book provides a comprehensive overview of psychometrics as well as a discussion of several practical aspects of the topic. Furthermore, important historical and ethical issues are presented. We summarize them in this subsection.

Generally, psychological assessment has diverse goals. Tests can potentially aim at recruiting the ideal candidates for a job or to create equality in educational settings by identifying learning disorders. Another controversial function is the goal of using psychometric profiling to build micro-targeted advertising to influence voting habits in democratic elections [3].

The roots of psychometrics reach back long before Darwin's famous publications *On the Origin of Species* and *The Descent of Man*. Talent was assumed to be a divine gift that depends on the mere judgment and plan of God [4]. However, Darwin's discovery of evolution had a great impact on the human sciences and launched a scientific project with the goal of revealing the impact of nature on human beings. Ever since, a key ambition has been the goal of measuring individual intelligence. "Intelligence is not education but educability. It was perceived as being part of a person's make-up, rather than socially determined, and by implication their genetic make-up. Intelligence when defined in this way is necessarily genetic in origin" [4, p. 8]. Thus, "for socio-biologists, intelligence test scores reflect more than the mere ability to solve problems: they are related to Darwin's concepts of 'survival of the fittest'. And fitness tends to be perceived in terms of images of human perfection. [...] Intelligence viewed from this perspective appears to be a general quality reflecting the person's moral and human worth, and has been unashamedly related to ethnic differences" [4, p. 16].

The attempts to find the common denominator of intelligence in the genetic make-up of an individual led scientists to the field of eugenics. The central hypothesis of eugenics is degeneration. A given population is degenerating if organisms with undesirable characteristics reproduce more quickly than the population with desirable characteristics. Based on this view, eugenicists stated that humans, "by caring for the sick and 'unfit', are undergoing dysgenic degeneration, and that scientists should become involved in damage limitation" [4, p. 10]. Eugenicists and their goals of selectively breeding humans were based on the concepts of the theory of evolution. "The intelligence testing movement at the beginning of the 20th century was not simply like Nazism in its racist aspects – it was its ideological progenitor." These "[...] ideas entered into the evolutionary theory and were used to dress-up dubious political beliefs in an attempt to give them a pseudo-scientific respectability" [4, p. 17]. After the Second World War and the racist crimes of the Nazis, eugenics was shunned by society. In today's world, the topic of racism is more sensitive, and it is safe to say that the originators of psychometrics did not share this sensitivity.

The sheer endless drive of scientists to find and define measures for intelligence led to the development of statistical methods, including correlation, normalization, standard deviation and factor analysis, which can all be used as methods for psychological assessment. It led further to the development of sets of items that are used for testing and can be compared to each other. These methods have been the foundation of standardized testing, influencing academic and career assessments on a daily basis. Therefore, the function of testing is determining its use, and this function derives from the need in any society to select and assess individuals within it. Given that selection and assessments exist, it is important that they be carried out as properly as possible and that they be studied and understood.

Psychometrics can be defined as the scientific process of selecting and evaluating human beings. But in the modern age we must realize that the ethics, ideology and politics of these selections and assessments are integral parts of psychometrics, as well as statistics and psychology. This concern arises in particular because "any science promoting selection is also by default dealing with rejection, and is therefore intrinsically political" [4, p. 25].

1.2.2 Psychometric schools and testing

In general, there are two schools within psychometrics: the trait school and the functional school [4].

The tests of the functional school are built in a linear way, which means that content areas on the *x*-axis are mapped against different levels of manifestations on the *y*-axis. Content areas can be political geography, for example, and manifestations are, for example, ratings on a scale from 1, bad, to 4, very good. An item in this case is basically the question that leads to the manifestation/answer of the content area function. Functional tests are commonly used for the assessment of job applicants or for the selection of candidates for programmes in higher education. The function changes depending on the goal of assessment.

The tests of the trait school, however, try to separate themselves from a purely goal-driven approach by generalizing answers into notions of human intellect and personality. This leads to the belief that personality types are not binary and exclusive, but an individual's personality is rather a mix of many traits and between each trait's extremes [4]. The most fascinating difference between the functional school and the trait school is therefore the attempt of the trait school to find the degrees of manifestations in personality types, while the functional school assesses the suitability of an individual for a given task.

Our study examines the trait school's implications for the assessments of personality traits. We investigate an individual's behaviour correlating with a larger population with respect to that individual's personality traits and the influence of personality traits on the decisions made by the individual.

A serious problem of testing in both schools is the theory of true scores. It is assumed that an observed score is a sum of the true score and an added error. This error can be based on biases. Bias can be caused by the construction of the test questions. Item biases are rather simple to identify. A test in the USA might be formulated with dollars and cents and would therefore not be appropriate for use in the UK. Linguistic forms of item bias are the most common ones [4]. Another bias is item offensiveness. Offensive items include racism and sexism. Intrinsic test bias exists when the test itself is constructed for a certain group and does not give adequate chances to a group that the test was not constructed for. A simple example is a native English-speaking group taking a test that was made for native speakers, but non-native speakers have to take the same test. Extrinsic test bias is found when there are actual differences between the social standings of the two mentioned groups in the example of the intrinsic test bias. Certain biases are regulated by law, for example racial bias. For instance, in Germany, it is not allowed to select candidates based on their ethnicity. Intrinsic test biases can be regulated by positive discrimination; it is much harder to regulate extrinsic test biases.

Biases are ubiquitous. Facebook, which we will use as our data source in this study, can be biased too. If argued from a statistical point of view, measured scores could be subject to statistical bias, for example when recommendation of content is built upon things the users liked already within their so-called bubble [7]. Also, item bias based on linguistic differences raises a problem.

Differential item functioning (DIF) tests analyse the deviations of answers within and between groups. DIF tests can therefore provide information about the existing intrinsic bias of a test, meaning that DIF tests point to differences between test takers. DIF tests help to explain differences, for example between cultures and sexes.

Finally, it is important to construct a test based on characteristics that help to make measures and items comparable. To achieve this goal, not only is the true score relevant, so too are the reliability, validity, potential standardization and normalization of a test. Constructing a functional test seems to be fairly straightforward. There is a clear goal to achieve, and item sets can be directed towards that goal. Designing a trait-based test, however, seems to be much more difficult. Not only is it important to define measurable personality traits, but the definitions of the traits and the selection of the items to serve the test's purpose have a lot of potential for biases.

1.2.3 The Big Five traits and politics

One of the biggest challenges of psychometric trait analysis, as described in the previous section, is to define the personality traits to be measured. "The origins of trait theory can be traced back to the development of the IQ testing movement, particularly to the work of Galton and Spearman. From the perspective of trait theory, variation in personality is viewed as continuous, i.e., for a specific personality, characteristics vary along a continuum. The advantage of a trait approach is that a person can be described according to the extent to which he or she shows a particular set of characteristics" [4, p. 150].

The definition of the term personality is still very much debated. An encompassing definition of personality does not exist, nor is one ever likely to emerge. Each definition is based on a different theory that is trying to explain human behaviour in a certain context and therefore contributes to a better understanding of what personality is. In the context of psychological testing, personality can be defined as an individual's unique constellation of psychological traits and states [8].

In the more specific context of micro-targeting of individuals based on psychometrics within social media, we extend the previously mentioned definition as follows:

Psychometric micro-targeting is the adaptation of content (pictures, videos, sounds and texts) based on an individual's unique constellation of psychological traits and states, to trigger certain favourable (to the content creator) actions of the content receiver.

A common approach to measuring and defining personality is to use factor analysis [9]. Factor analysis is a vectorial method that is based on correlations of manifestations of individuals' responses to items, using vectors to describe influencing factors that lead with a certain multiple to the measured output. Factors in psychometrics are basically the hidden influencers of human decisions. It is important to have as few factors as possible and to have common factors throughout psychometrics for the description of human behaviours, to make measured outcomes more comparable.

Progress in the field of psychometrics was made possible by the adoption of the Big Five model as the unifying force of the field of personality.

Donald Winslow Fiske "was the first who noticed that with five factors it was possible to obtain similar factor definitions when different assessment techniques, such as self-ratings, peer-ratings and observer ratings, were used" [4, p. 166]. The Big Five traits are extraversion, agreeableness, conscientiousness, emotional stability/neuroticism and openness to experience [5]. A short overview of these traits is presented in Table 1.1.

Tab. 1.1: The Big Five traits [5, p. 267]

Trait	Definition
Extraversion	...energetic approach towards the social and material world
Agreeableness	Contrasts a prosocial and communal orientation towards others with antagonism...
Conscientiousness	...socially prescribed impulse control that facilitates task- and goal-directed behaviour...
Emotional stability	Contrasts...even-temperedness with negative emotionality...
Openness to experience	...the breadth, depth, originality and complexity of an individual's mental and experiential life

The conclusive definition of the Big Five model is an argued standard within the psychometric community. Four main reasons support the acceptance of the Big Five model. The first one is that the five traits have high stability. Secondly, the traits are compatible with a wide range of psychological theories. Thirdly, the five traits occur in many different cultures. Finally, the five traits have a biological basis [4, 10].

Obviously, there are various options with which micro-targeting, applying the Big Five traits, could take place. One might think of opportunities for companies in the field of marketing, for example. However, in this chapter, we consider the context of politics, especially because of its current importance (refer to the examples of election campaigns mentioned in the introductory section).

Generally, making use of the Big Five traits when analysing elections is not a new idea. To name just a few studies, Vecchione et al. (2011) elaborated for Italy, Spain, Germany, Greece and Poland that the Big Five were linked to party preference. The traits have substantial effects on voting, while socio-demographic characteristics (gender, age, income and educational level) had less influence. The openness trait has been shown to be the most generalizable predictor of party preference across the examined countries. Conscientiousness was also a valid predictor, but its effect was less robust and replicable [11].

Dennison (2015) draws a similar picture for the 2015 general election in the UK: "Undoubtedly the two most consistently found relationships are the positive effect of conscientiousness on right-wing voting and the positive effect of openness to experience on left-wing voting" [12]. The rationales behind this might be that very conscientious people, for which socially prescribed norms and rules are more important, are rather conservative. In contrast, open-minded people could be characterized as open to unconventional and even unorthodox political approaches, which is generally more associated with left-wing parties. Dennison (2015) also emphasizes that emotional instability tends to have an influence in favour of left-wing parties: "Emotionally unstable individuals are more anxious about their economic future, more desirous of state control, and are less likely to view the status quo in positive terms – all of which theoretically increases the chance of left-wing attitudes" [12]. Figure 1.2 shows the analyses of Dennison (2015) for all Big Five traits. He applies *z-scores*, which are a numerical

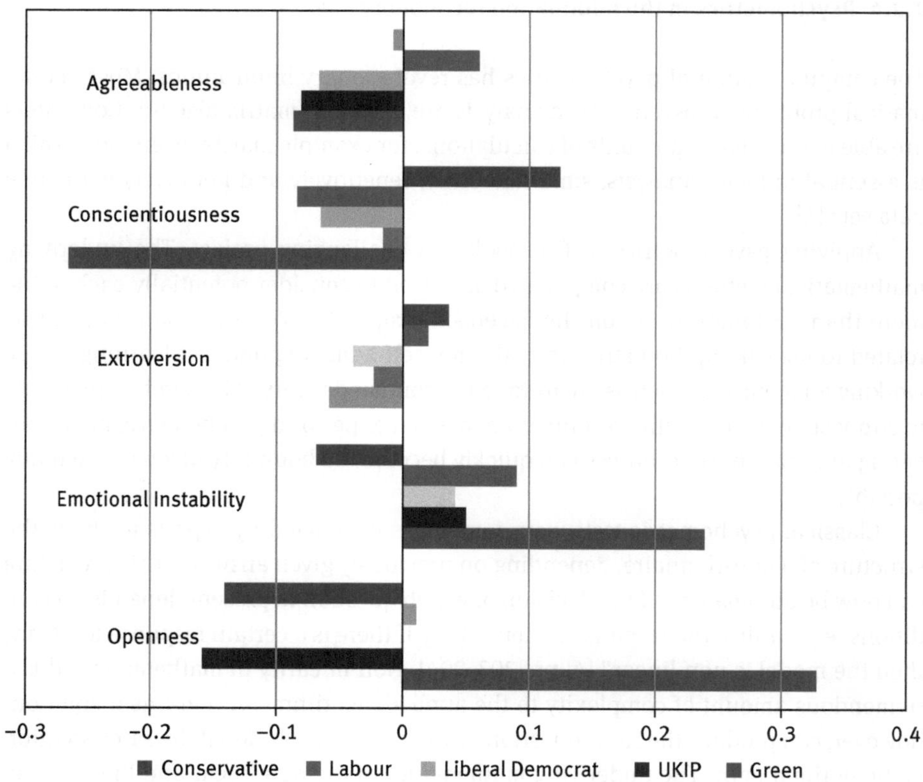

Fig. 1.2: Personality traits and party choice in 2015 in UK [12]

measure of a value's relationship to the mean in a group of values. For example, if a z-score is 0, the value is identical to the mean value. A positive z-score indicates the value is above the mean and a negative score indicates it is below the mean. Considering the case of openness, voters of the right-wing UK Independence Party (UKIP) have the highest negative value, which means that they are the most closed ones. In contrast, Green voters have the highest positive openness value, which indicates that they are the most open-minded people in this sample.

Based on these examples, we conclude that there seems to be a link between individual psychometric traits and voting behaviour. The aforementioned studies mainly refer to self-assessment of individuals, meaning that the trait scores are derived from questions asked in surveys. We believe that in times when social media services are one of the main communication channels, these Big Five traits could be extracted from social media behaviour. We outline some general remarks on psychometrics in the digital age in the following subsection before presenting our conceptual approach in Section 1.3.

1.2.4 Psychometrics in the information technology age

The computerization of psychometrics has revolutionary implications. Most mathematical problems of psychometric analysis are based on matrix algebra. Computers are able to do massive amounts of calculations, for example matrix inversions, which is essential to factor analysis, simultaneously, repetitively and iteratively with large data sets [4].

Applying psychometrics to Facebook is a challenging project. The underlying mathematical methods are complicated and need to consider potentially each of the more than 1.3 billion users on the Facebook Graph. The collected data need to be related to lexical psychometrics in real time, self-adjusting and self-learning, while working for a purpose such as micro-targeted content delivery. It is therefore plausible to suppose that substantial resources are needed to perform psychometric micro-targeting and that those resources can quickly become the boundary of what is actually possible.

Classical psychometric tests were sometimes conducted by experts to shape the structure of a questionnaire, depending on previously given answers. This procedure can now be automated. “If the decision of which question to present depends on conditions, e.g., utilize the response to item x only if there is a certain response to item y, then the model is non-linear” [4, pp. 203–204]. Non-linearity in mathematics adds a tremendous amount of complexity to the applied algorithms. A non-linear solution, however, can produce the same questionnaire as a linear solution if the same solution is the optimal path. The underlying mathematics of the non-linear and linear structures are equal in terms of statistics; the complexity derives from the decision-making abilities of the non-linear system.

"A neural network trained to recognize the possibility of diverse pathways to the same standards of excellence could potentially outperform any paradigm from classical psychometrics that was by its nature restricted to linear prediction" [4, p. 205]. Thus, it is possible to use a Bayesian approach that always chooses the next content to be shown in a way that maximizes the probability that the assessed individual will make the preferred decision. If the assessed individual makes the preferred, and therefore predicted, decision, the algorithm adjusts itself with a certain success factor. If the assessed individual denies the decision, the algorithm will understand the assessed individual better and adapt the presented content.

While neural network programs can learn from experience to make excellent behavioural predictions, the internal procedures they follow are often much too complicated for any human to understand. A characteristic of non-neural psychometrics is that the models try to identify latent traits so as to correspondingly adjust the personality traits of the assessed individual. The neural psychometric approach does not rely on latent traits; its algorithm constantly screens patterns and changes the underlying assumptions anytime it succeeds. The predictions are purely actuarial. A good neural network has strong predictive powers; the disadvantage, though, is that there is hardly a human being that can understand how the predictions are made. "Unlike expert systems, neural networks include no explicit rules and have no justification other than their success in prediction" [4, p. 207].

A machine that is able to predict real-world decision outcomes of individuals has tremendous value. However, even an imperfect solution that describes psychometric profiles is interesting to "personnel and credit agencies, the insurance and marketing industry, social security, the police and intelligence services" [4, p. 198]. That is why it is important to take special care during data collection and analysis. Computer systems are also able to generate reports. "Many computerized testing or scoring programs no longer report mere numbers, to be interpreted by experts, but are able to produce narrative reports in a form that is suitable for the respondent or other end users. Where the test is a profile battery, the computer is able to identify extremes, to interpret these in the light of other subscale scores, and to make recommendations" [4, p. 201].

1.3 Methodological framework

In this section, the underlying statistical methods and a technical framework to integrate all aspects into one system that is able to calculate predicted probabilities for personal traits are presented.

First the latent Dirichlet allocation (LDA) [13, 14], which is used intopic modelling, is introduced.

Then the implementation of the algorithms in the programming language R is described. The code and a detailed description of the implemented algorithms as well as alternative approaches can be found in *Mining Big Data to Extract Patterns and Predict Real-Life Outcomes* by Kosinski et al. (2016) [15] and on the related project website http://mypersonality.org [16]. The data available from the myPersonality project are used to calculate the prediction models, which serve as the foundation for later estimations.

Then a set of software components is presented that can be combined to build a working prediction environment. Because there is a multitude of possible webserver software and corresponding plug-ins, the goal is to show one setting in detail and briefly introduce alternatives that can be used to customize the system and adjust it to the needs and prerequisites of different technological landscapes.

Finally, we describe how Facebook data can be integrated in our model while presenting corresponding coding examples.

1.3.1 Latent Dirichlet allocation

The LDA is used for topic modelling [13]. A topic is a collection of words that have different probabilities of appearing in passages discussing the topic. If the topic producing the words in a collection is known, it is possible to guess and assign new words that relate to the given topic. The way this is done is by considering the number of times the word occurs in the discussion of the topic and how common the topic is in the rest of the documents. Human memory capacities are limited. While a human being can understand latent structures in a limited amount of texts, the LDA refers to a large scale of texts that a human being would not be able to process in a relatively short period of time [14, 17].

One way to explain the underlying mathematics is to visualize it in a simple and reduced Bayesian statistical formula. Figure 1.3 depicts a linear approach in which for each topic Z, the frequency of a word type W in the topic Z is multiplied by the number of other words in document D that already belong to Z. The result represents the probability that the word W came from the topic Z. Depending on which one has the highest probability, the word will be sorted into one of the given topics. This expression is a basic Bayesian formula that describes the conditional probability that a word W belongs to topic Z [17].

$$P(Z|W,D) = \frac{\#\ of\ word\ W\ in\ topic\ Z + \beta_w}{total\ tokens\ in\ Z + \beta} \cdot (\#\ words\ in\ D\ that\ belong\ to\ Z + \alpha)$$

Fig. 1.3: Simplified (linear) LDA algorithm [17]

The connection between a word and a topic Z influences the total probability *a priori*. Based on this, the machine learns by applying topics and the topic-constructing words on more documents. The complexity of the LDA therefore grows when the Bayesian model is translated into vectors, resulting in the need to define functions for the probabilistic distributions relating to the prior assumptions, the indicator function and the a-priori probability. Further complexity is added by expanding a usually two-level Bayesian model into a three-level hierarchical Bayesian model, in which each word is a mixture of underlying topics and each topic is a mixture over an underlying set of topic probabilities. The goal is to assure that the essential statistical relationships between each layer are preserved [13]. The preservation of statistical relationships potentially allows a set of applications that can enhance the model with external information.

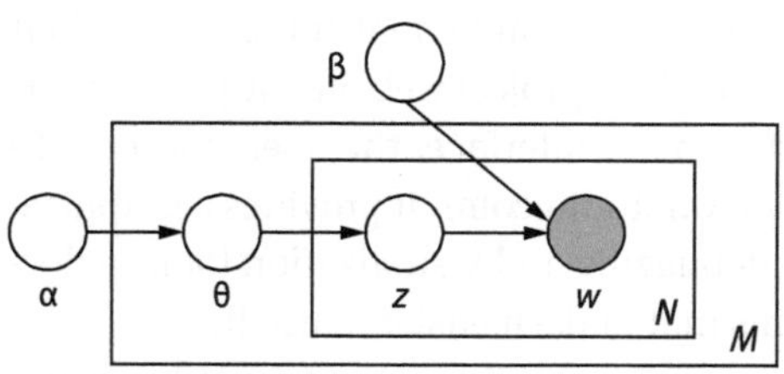

Fig. 1.4: Graphical model representation of LDA [13, p. 997]

Figure 1.4 describes the variables defining the LDA function in a graphical representation. A word is described as w and defined to be an item from a vocabulary indexed as a vector and represented by a unit vector relating to the vocabulary vector with a single component equal to one and all other components equal to zero. A document is a sequence of N words. A corpus is a collection of M documents. The variables α and β are corpus-level parameters, assumed to be sampled once they are in the process of generating a corpus. The variable θ is a document-level variable, sampled once per document. As indicated in Figure 1.4, there are three levels of the LDA representation: the word, the document and the corpus. The LDA algorithm chooses N words based on a Poisson distribution. Each of the N words is pointed towards a topic that has been chosen by the application of a Multinomial Dirichlet application on θ describing the probability distribution, and a Bayesian multinomial probability conditioned on the topic for each word [13].

Identifying the probability densities based on the multinomial distribution of words relating to topics helps to cluster the relevant words in topics and ideally to generate a "term-by-document matrix [. . .] that reduces documents of arbitrary length to fixed-length list of numbers" [13, p. 994]. These numbers therefore help to reduce the dimensionality of matrices to relevant "keywords" and "topics". Concerning the goal of the Big Five personality assessment, it is plausible to expect $k = 5$ topics. All words in the given Facebook information are clustered iteratively. Words that are not able to be categorized into a topic with a certain amount of probability can be deleted, reducing the complexity of the initial user-like matrix that our study uses based on the myPersonality project to execute the algorithms.

The LDA allows one to build topics. The resulting topic databases contain words that can be correlated against psychological lexica like the *Linguistic Inquiry and Word Count*Linguistic Inquiry and Word Count (LIWC) [18]. The correlation coefficients can be used to build Big Five personality trait models in an additional database.

1.3.2 Statistical programming

As the first step, R, a language for statistical computing, should be installed; this will enable user to follow the presented instructions and implement the code on their own. A current version can be downloaded from the R project website: https://www.r-project.org/. As R itself only provides a rudimentary interface, the open-source software RStudio can be installed from https://www.rstudio.com/. It provides a graphical user interface that integrates a code editor, debugging and visualization tools, a documentation browser and additional functions that make it easier to use R.

To begin with, the data sets provided by myPersonality project must be downloaded. These data were collected by the myPersonality Facebook application "that allowed users to take real psychometric tests, and allowed [us] to record (with consent!) their psychological and Facebook profiles" [16]. In the data sets, the scores of the psychometric tests, the records of the users' Facebook profiles and item-level data are available. After saving the .csv files to the R project folder, they can be loaded into the data environment.

```
users <- read.csv("users.csv")
likes <- read.csv("likes.csv")
ul <- read.csv("users-likes.csv")
```

Now, *users* contains the anonymized userid, gender, age, political views and scores for the five personality traits. In *likes*, the name and ID of all likes made by users are stored. To be able to connect both of these together, *ul* specifies which user made which like by providing pairs of userid and likeid. The next goal is to create a digital user footprint matrix, a matrix where each like is presented as a column, each user is presented as a row, and the entries contain the value 1 or 0 to indicate the existence or non-existence of the corresponding like. Therefore, in the *ul* object, the userid and likeid are matched with their respective rows in *users* and *likes*. Afterwards, the *sparseMatrix()* function is used to create the desired footprint matrix *M*, where the row names are set to the corresponding userids and the column names are set to the corresponding like names.

```
ul$user_row <- match(ul$userid,users$userid)
ul$like_row <- match(ul$likeid,likes$likeid)

M <- sparseMatrix(i=ul$user_row, j=ul$like_row, x=1)

rownames(M) <- users$userid
colnames(M) <- likes$name
```

As the resulting matrix contains a large number of unique likes, the density is comparatively low, and the matrix should be trimmed. There are no correct thresholds. In general, users with few likes or likes made by only a few users do not significantly contribute to the model's explanatory power but require more computing time. Using relatively high thresholds, all users with less than 50 likes and all likes that were made by less than 150 users are removed. Users that are removed from *M* must also be removed from *users*.

```
repeat {
  i <- sum(dim(M))
  M <- M[rowSums(M) >= 50, colSums(M) >= 150]
  if (sum(dim(M)) == i) break
}

users <- users[match(rownames(M), users$userid),]
```

After these steps, a user footprint matrix is available and can be used to find patterns or, stated better, extract patterns and build prediction models. In the field of LDA, clusters are called topics, as the method originates from language processing, where it is used to identify topics [13]. Using the created sparse matrix *M* as input for the *LDA()* function from the *topicmodel* library, an MDA model is computed. To ensure that the same results are obtained as presented in this chapter, R's random number generator must be set to 68. *Mlda* now contains 50 clusters that were identified based on the user footprint matrix.

```
Mlda <- LDA(M, control=list(alpha=1, delta=.1, seed=68), k=50,
method="Gibbs")
```

To better understand the LDA algorithm and its results, it is helpful to run the preceding computation with $k = 50$, meaning the algorithm will extract five clusters from the data set, and take a closer look at the results. While this is helpful to understand the approach, having more clusters usually increases the precision of the derived prediction model. A detailed discussion of the influence of varying k can be found in Kosinski et al. (2016) [15].

Several strategies can be applied to interpret clusters. Exploring footprints that are associated with the clusters is the first step. The second step is to identify the relationships between dimensions and clusters with user information like demographic information, psychological traits, collected questionnaires or a lexical database like the LIWC [18].

```
gamma <- Mlda@gamma
cor_lda <- cor(gamma, users[,-1], use = "pairwise")

x<-round(cor_lda,2)
y<-melt(x)

colnames(y)<-c("LDA", "Trait", "r")
qplot(x=LDA, y=Trait, data=y, fill=r, geom="tile") +
  scale_fill_gradient2(limits=range(x), breaks=c(min(x), 0, max(x)))+
  theme(axis.text=element_text(size=12),
        axis.title=element_text(size=14,face="bold"),
        panel.background = element_rect(fill='white', colour='white'))+
  labs(x=expression('LDA'[rot]), y=NULL)
```

Figure 1.5 shows a heat map describing the user scores on LDA clusters of the R object gamma's correlation to psychodemographic traits; it reveals that gender, age, political views and the personality trait of openness ("ope") are the ones most strongly correlated with the extracted LDA clusters.

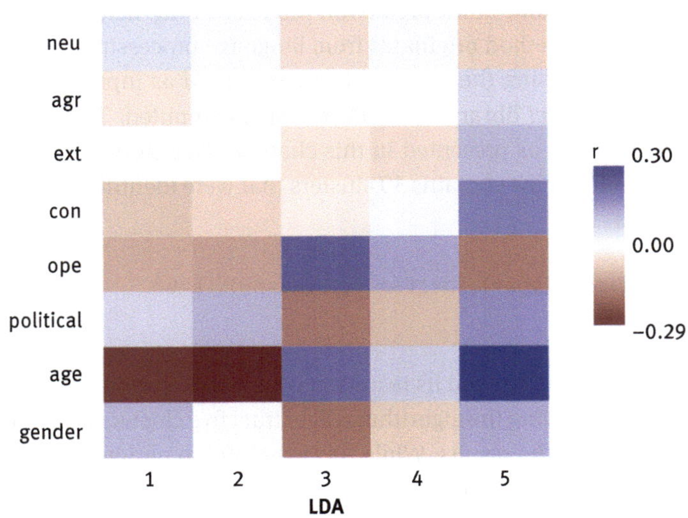

Fig. 1.5: Heat map showing correlations between users' membership in LDA clusters and their traits [15, p. 500]

To estimate the predictors for an extracted topic, the conditional distribution of the proportions given the data and the prior distribution must be calculated, which is done using the *posterior()* function. The calculated posterior distribution for the topic is then stored as a data frame in *predictors*.

```
predictors <- as.data.frame(posterior(Mlda, M,
control = list(alpha = 1, delta = .1))$topics)
```

Having the predictors, it is now possible to create linear regression models that specify the factors with which to multiply the users' topic memberships when calculating the predicted probabilities for variables. The *glm()* function is used to calculate generalized linear regression models, which are then stored in *fit*. For the calculation, *var* must be replaced by one of the abbreviations of the variables (gender, age, political, ope, con, ext, agr), or it can be wrapped in a loop to calculate the models for all variables. To estimate the predicted probabilities for user traits, the *predict()* function needs a fitted model, in this case a linear regression model, and the predictors as input.

```
preds<-list()
for (var in vars){
  preds[[var]]<-rep(NA, n = nrow(users))
  # check if the variable is dichotomous
  if (length(unique(na.omit(users[,var]))) ==2) {
    fit <- glm(users[,var]~., data = predictors, family = "binomial")
    preds[[var]] <- predict(fit, predictors, type = "response")
  } else {
    fit <- glm(users[,var]~., data = predictors)
    preds[[var]] <- predict(fit, predictors)
  }
}
```

To acquire reliable information about the quality of an estimation, split the data set into a training set and a testing set. Then the fitted model must be trained independently from the users for which the predicted probabilities are to be calculated. For a detailed description of this process, we refer the reader to the original paper [15].

After these steps, the predicted probabilities for all variables are now stored in *preds*. To obtain probabilities for a specific variable for a specific user, the variable abbreviation and userid must be specified:

```
preds["ope"][preds$userid=="fc56bb60bfe7044f28dbdd5a81097709"]
```

1.3.3 Infrastructure

In this subsection, we introduce a possible deployment model that describes the devices and execution environments in each tier. The purpose is to give an example of the configuration of and the possible alternatives for the server infrastructure, including all devices, execution environments, processing elements and connectors of the system's parts. The prototype system consists of a server infrastructure with a client tier, a web tier, a business tier and a database tier.

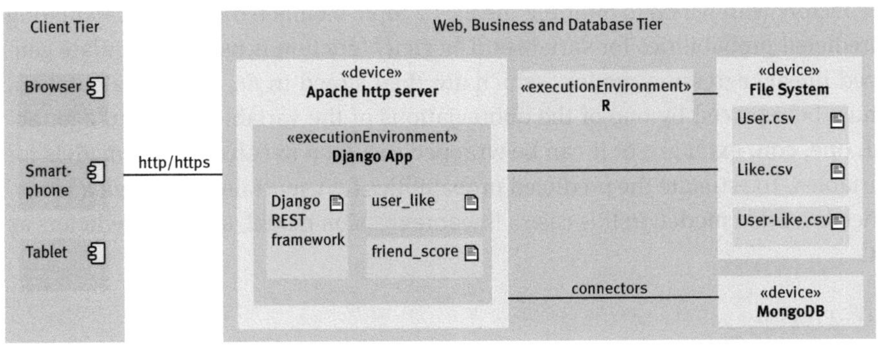

Fig. 1.6: Model of conceptual deployment of prototype

Figure 1.6 presents the deployment model of an environment that can collect, evaluate and present information that characterizes an individual's personality traits in relation to her Facebook likes. As seen in Figure 1.6, the client tier can consist of various devices. It is recommended to use a framework that can provide well-designed interfaces fitting the requesting device (e.g., browser, smartphone or tablet). The information is encoded in JSON to be passed between the different elements within the system. Initially, a device sends a request and the included information is pushed to the web, business and database tiers. Facebook, for example, offers different API calls for user information for browser, smartphone and tablet devices and the functions that are implemented in their respective operating systems.

Once received by the Rest API of the Django Rest Framework, the information is pushed through a Python script that persists and checks the data for accurate formatting. If the format of the incoming data does not match the expectation of the script, an error message is sent back. This step is necessary to be able to react to possible changes in the Facebook API without having to refactor the underlying business logic. To run the Django Rest Framework, an Apache Server is used. The processed user data are stored to the file system and added to the user and user-like .csv files. The R execution environment has access to the .csv files, allowing the system to run the required algorithms, which are implemented as R scripts. The results can be written to the file system, for example using the R function *write.csv()*. Since the file system should be accessible by other elements of the system, the results can be used to generate output concordant with the user's personal traits that were predicted by the R environment.

1.3.3.1 Technologies

The technologies used and implemented in the prototype system were chosen according to various functional and non-functional requirements. This subsection gives an overview of the implemented technologies, which will be extended by potential future developments and suggestions for future research.

Python was chosen as the preferred script language for data analysis and as the foundation of the web framework that manages the Rest API. As one of the most widely used languages for scientific applications, Python has a reputation of being easy to learn, is minimalistic and is free of charge. There is a growing international community of scientists and professionals who contribute libraries and frameworks on a regular basis that help build efficient scientific applications. Libraries such as SciPy and the topicmodels package can help to reach a higher level of automation with the LDA algorithm [15]. However, the used LDA algorithm is built in R. Python is only applied within the data approval script that lies within the Django framework.

The Django framework is a Python framework that helps developers build and use fast, secure and scalable interface solutions for the external communications of a server. This helps developers focus on the web application, rather than the development of the basic infrastructure.

Apache Web Server is an open-source HTTP server for operating systems including *NIX and Windows. It is one of the most used web servers. Basically, the server is in charge of delivering all kinds of content. Its functionalities can be extended with modules. The source code of the Django framework is interpreted by the Apache Web Server and then delivered to the client.

R is a programming language for statistical computing. Its extensive libraries and core functionalities make R one of the programming languages most used by statisticians. It is open source, and a large community of mathematicians and software developers has formed around it. Therefore, there are many plug-ins, APIs, modules and manuals, making it a good fit for almost every environment without the need for a large financial investment.

JSON is a lightweight data-interchange format and has become a standard for the transportation of data between different architectures and software interfaces. It is built on two structures: firstly, a collection of value pairs, realized as objects, and secondly, an ordered list of values, which can be understood as vectors. From the perspective of algebraic mathematics, relational databases form large-scale matrices, and JSON objects form vectors.

1.3.4 Facebook integration

Facebook's Graph API Explorer (https://developers.facebook.com/tools/explorer/) is an amazing tool for testing and understanding the functionality and possibilities that the Graph API offers. As shown in Figure 1.7, one can easily request likes of a user, who must consent to this action, to be reported as JSON output. In this example, the likes from the account of one of the authors is pulled by *GET /v2.9/me?fields=likes*, which means that the Graph API Version 2.9 is used, to query the object *me* for the values from all *fields* of the type *like*.

Fig. 1.7: Example of Facebook Graph Explorer tool

Testing commands is an essential step in designing Facebook logins for applications that pull user data in order to drive the analysis presented in this chapter. Online surveys or questionnaires can be integrated directly in Facebook as HTML code, including a Facebook login button that allows researchers to pull and record Facebook user information with the consent of the user [15]. For the purpose of personality trait prediction, the JSON information received from the Graph API can then be further preprocessed and formatted and then passed to the R execution environment. An example of a software development kit (SDK) that uses Facebook social Graph APIs is the React Native SDK (https://developers.facebook.com/docs/react-native). React Native is an SDK that was developed by Facebook to create native applications for iOS and Android mobile operating systems with congruent JavaScript code. React Native saves developers time by publishing one code for two operating systems. The React Native SDK can be used to pull profile information from mobile application sign-ups. This step is especially interesting for businesses, since sign-ups with Facebook could help them to collect a large amount of essential information about their customers.

The first step in accessing the Graph API is to define a function that allows users to enter login information and to authorize the application to grant access permission to the user in case of a successful login.

```
const FBSDK = require('react-native-fbsdk');
const {
  LoginButton,
} = FBSDK;

var Login = React.createClass({
  render: function() {
    return (
      <View>
        <LoginButton
          publishPermissions={["publish_actions"]}
          onLoginFinished={
            (error, result) => {
              if (error) {
                alert("Login failed with error: " + result.error);
              } else if (result.isCancelled) {
                alert("Login was cancelled");
              } else {
                alert("Login was successful with permissions: "
                + result.grantedPermissions)
              }
            }
          }
          onLogoutFinished={() => alert("User logged out")}/>
      </View>
    );
  }
});
```

Afterwards, the login button is included in the rendered view of the application, so that the user can actually interact with the application.

```
class MyApplication extends Component {
  render() {
    return (
      <View>
        <Text>Welcome to the Facebook SDK for React Native!</Text>
        <Login />
      </View>
    );
  }
}
```

The final step is to remodel the login button. With respect to granting permission, the access token can be used to request access to the data of the user and then parse it as JSON. If the permission is not granted, the function throws an error.

```
<LoginButton
  publishPermissions={['publish_actions']}
  readPermissions={['public_profile']}
  onLoginFinished={
    (error, result) => {
      if (error) {
        console.log('login has error: ', result.error)
      } else if (result.isCancelled) {
        console.log('login is cancelled.')
      } else {
        AccessToken.getCurrentAccessToken().then((data) => {
          const { accessToken } = data
          initUser(accessToken)
        })
      }
    }
  }
  onLogoutFinished={logout} />

// Definition of the initUser() function used above
initUser(token) {
  fetch('https://graph.facebook.com/v2.5/me?
  fields=email,name,friends&access_token=' + token)
  .then((response) => response.json())
  .then((json) => {
    // Some user object has been set up, build it here
    user.name = json.name
    user.id = json.id
    user.user_friends = json.friends
    user.email = json.email
    user.username = json.name
    user.loading = false
    user.loggedIn = true
    user.avatar = setAvatar(json.id)
  })
  .catch(() => {
    reject('ERROR: No data received from Graph API')
  })
}
```

1.4 Testing the framework

By now, the theoretical, methodological and technological foundations have been laid. Two cases that demonstrate the actual application will now be presented. In addition, we will discuss the link with the political context, i.e., what political preferences can be assumed and how parties could make use of this information.

1.4.1 Use case 1: I am what I like

Assuming that the likes of a new Facebook user who has not been included in the original data are available, the goal is to predict the probabilities of this user's personality traits. For this example, a user called Peter was created. A script was then used to randomly assign between 300 and 400 likes from the original likes data set to Peter, resulting in a total of 347 likes. For example, some of the assigned likes are

- http://www.facebook.com/pages/West-Coast-Port-Shutdown/100856866698826
- Telling Rush Limbaugh he's Full of Crap (by Leftake.com)
- Super Nintendo Entertainment System

As the first step, each new user's information must be included as a digital footprint in the sparse matrix *M_user*. Therefore, the new user's information is added to users.csv and user-likes. before starting to process the data. To save computation time, it is further recommended to take a look at the newly created sparse matrix and extract a partial sparse matrix that contains only new users. *tail(M_user)* shows the end of the matrix and *M_user[startrow:endrow,]* selects only the specified range of rows. *dim(M_user)* is helpful for determining the required number of rows.

To calculate the predicted probability for a new user, the fitted model that was trained as described in Section 1.3.2 is used. *user$var~* specifies that the variable *var* is estimated using all independent variables. Additionally, predictors for the new user(s) are created. Then both are used as input for *predict()*.

```
M_part <- M_user[110729:110730,]
preds_Part <- as.data.frame(posterior(Mlda, M_part,
control = list(alpha = 1, delta = .1))$topics)

fit <- glm(user$var ~ ., predictors)
prediction <- predict(fit, preds_Part)
```

After running the calculations, the following values are predicted for Peter:

gender	political	age	ope	con	ext	agr	neu
0.014	0.033	38.07	1.087	-0.401	-0.942	-0.257	-0.143

Based on the randomly assigned likes, the model estimated Peter to be a 38-year-old male Democrat. Taking a closer look at his psychodemographic traits, he can be classified as a relatively open-minded person, leading to the assumption that his political views are left wing. Even though the relationship is not yet as well established in the literature, his relatively low emotional stability (neuroticism) indicates a left-wing orientation as well. For the interpretation of the predicted values, it is helpful to use the previously presented heat map function as well as *summary(users)* to get an understanding of the possible range of values for each trait and its distributions.

Given the derived political preferences, parties could make use of this information. Considering a scenario in which a left-wing party is able to identify Peter's political orientation, the party could try to strengthen his political preferences, for example by providing studies that underline the need for social equity. On the other hand, in a scenario in which a right-wing party knows about Peter's left-wing orientation, the party could try to provide information that emphasizes the need for more conservative actions, thereby trying to weaken Peter's current commitments. Besides the approach of strengthening those preferences of an individual that seem favourable, a party could be more direct and present information that stresses how the target's preferences align with those of the party, implying that a vote for this party would be a good choice.

Influencing a person based on individual analysis is not a new phenomenon. Micro-targeting is, for example, very popular and sophisticated when used for marketing purposes. All Internet users have experience with personalized pop-up advertisements that are based on website visits and other previously taken online actions. However, coming back to the example of Peter, there is a very high risk of fake news. In our context, this would mean that a political party uses intentionally misleading or even wrong information to influence Peter.

1.4.2 Use case 2: It started with a like

Besides the possibility of using users' likes to estimate their predicted user traits, one could start with a like that has been shown to be strongly correlated with a relevant trait and target all users with that specific like. Certainly, this approach is a very broadly based approach and will lead to the targeting of many ill-suited individuals. However, taking the constraints of computation time into consideration and, even more importantly, lacking user consent for the retrieval of these personal data, such as users' likes, this might turn out to be a feasible approach.

Again, this use case is based on extracted LDA clusters. For this description, $k = 5$ is used to extract five LDA clusters. While more clusters improve the precision when building predictive models, fewer clusters are easier to interpret and understand their commonalities. The LDA analyses used in this example are the same as those used to create the heat map in Section 1.3.2. Therefore, it is recommended to keep the heat map and the explanation of the clusters in mind while looking at the most strongly correlated likes shown here.

```
Mlda <- LDA(M, k = 5, control = list(alpha =10, delta = .1, seed = 68),
method =
beta <- exp(Mlda@beta)
top <- list()
for (i in 1:5) {
  f <- order(beta[i,])
  temp <- tail(f, n = 5)
  top[[i]]<-colnames(M)[temp]
}

print(top)
[[1]]
[1] "I hate it when i'm taking a drink and all the ice
attacks my face"
[2] "Telling inanimate objects to STAY when they look like
they're going to fall"
[3] "I feel stupid when I say what? a thousand times because
I can't hear"
[4] "i finally stop laughing... look back over at you and
start all over again"
[5] "Ok, If we get caught here's the story..."

[[2]]
[1] "Music"     "YouTube"   "Oreo"       "Starburst" "Skittles"

[[3]]
[1] "Music"               "The Daily Show"    "The Colbert Report"
[4] "Queen"              "The Beatles"

[[4]]
[1] "The Beatles"            "Red Hot Chili Peppers" "Nirvana"
[4] "Linkin Park"            "Family Guy"

[[5]]
[1] "Victoria's Secret" "The Hangover"      "Rihanna"
[4] "Eminem"            "Lil Wayne"
```

The strength of the correlation between the extracted LDA clusters and the likes contained within these clusters is stored in *Mlda@beta*. The list *top* is created and used to store the likes with the highest correlations, found in a *for* loop using *order()* to order the likes by the strength of their correlation and then *tail()* to get the end of the list.

The heat map shows that LDA cluster 3 is strongly positively correlated with openness. As openness tends to be a trait of more liberal/left-wing people, one could target Facebook users who like *The Beatles, Queen, ...* with suitable messages. On the other end, cluster 5 is strongly positively correlated with conscientiousness and, consonant with the theory, negatively with openness. This could imply that the users of *Lil Wayne, Eminem, ...* are more receptive to conservative/right-wing messages.

The potential use of this information is, however, the same as discussed in use case 1. Parties could present information that strengthens the traits associated with a certain like to all persons with that like or even customize an advertisement that incorporates the topic of the like and shows how the party itself views the topic and embodies the trait in question.

1.5 Discussion and conclusion

In this section we discuss the main contributions of our approach. In addition, its limitations, potential future developments and some concluding remarks are presented.

1.5.1 Contributions to research and practice

Our study makes three main contributions to research and to practice. First, it represents an easy and comprehensive introduction to Big Five psychometric trait assessment; second, it provides a conceptualization and prototyping of an infrastructure that supports a scalable solution for real-time data collection and evaluation; and third, it provides a proof of concept that psychometric traits can be actually used to predict real-life outcomes with a scalable infrastructure.

Our chapter gave a comprehensive introduction to computational psychometric trait analysis based on the Big Five model. It contains descriptions in terms of the conceptual frameworks of psychometric analysis, the meanings of the words in lexical analysis, and how to use Facebook as a research tool.

Furthermore, an instantiation of a prototype infrastructure was presented, with deployment diagrams and algorithmic methods connecting all the devices, execution environments, files and databases. This prototype describes what a scalable analytical solution for Big Five personality trait assessment could look like and has the potential for further automation and scalability. The prototype could be used by other researchers to set up a server environment for their own future research.

Finally, our chapter laid down a foundation for an objective understanding of computational psychometric trait analysis and can therefore be used as an argument to address issues in the field of privacy infringement and potential issues in the field of election fraud.

1.5.2 Limitations and suggestions for future research

One limitation of our study is its focus on Facebook as the primary data source. Several other social media services might be used to derive psychometric traits. However, we believe that our general approach could easily be adapted to other data sources.

The presented prototype is far from being optimized for speed and would require a fast computation environment to perform just-in-time calculations. In particular, the LDA algorithm and the training of the GLM model take a long time. Both algorithmic optimizations and an improvement of the technological implementation (e.g., in-memory storage of computed models) would be required to establish an efficient and feasible production system.

Having a large data set that combines psychodemographic traits with additional information, such as that provided by the myPersonality project, makes it possible to perform a variety of statistical analyses and prediction models. As the data set is static, it is not evolving, and newly created likes are not included. Given the volatility of today's technology and online trends in social media, important aspects and newly arising influence factors might not be included in the models.

The main goal of our chapter was to build an initial proof of concept to enable further research and development. Having built a prototype, it is important to point out three main suggestions.

Algorithmic suggestions. Scientists should pay attention to the methods used to build correlations between the LDA [14] clusters and the psychodemographic traits of users. While we used used a simple log-likelihood function, it is advisable to extend the method and build machine learning algorithms [19], ideally on neural networks, that are able to include user actions in the interactive re-ranking of trait scores [20]. When adding machine learning concepts to the assessment of personality traits for a more automated assessment of individuals [21], it is advisable to rethink the importance of the priors within the Bayesian statistics [22].

Architectural suggestions. The main architectural suggestions in Figure 1.8 are concerned with the elimination of the .csv file system currently being used in the prototype. A system based on .csv files is difficult and unreasonable to scale due to the limitations of access rights to such files. The next step in the development of the system would be to separate the database tier from the web and business tiers and use connectors from the web and business tiers to the database tier to push and pull data. MongoDB is the most widely used NoSQL database. NoSQL systems were developed to meet the requirements of building modern applications that create and process massive volumes of data. One strength of MongoDB is that it makes it easy to handle rapidly changing data types and therefore operate with structured, semi-structured, unstructured or polymorphic data. Code for MongoDB can be written to be agile, quickly pushed, and released multiple times a day. Relational databases were not built to manage the volume, variety and velocity of modern applications [23]. The MongoDB database tier would allow multiple users to access the contained data. It would also be easier to replicate databases or split the data into many different databases and servers. The separation between the business tier and the database tier helps in the use of tools like Hadoop and Spark to allow parallel computing and methods like map-reduce [24]. This separation would allow scalability in the infrastructure and would therefore solve the problems of data volume, velocity and volatility.

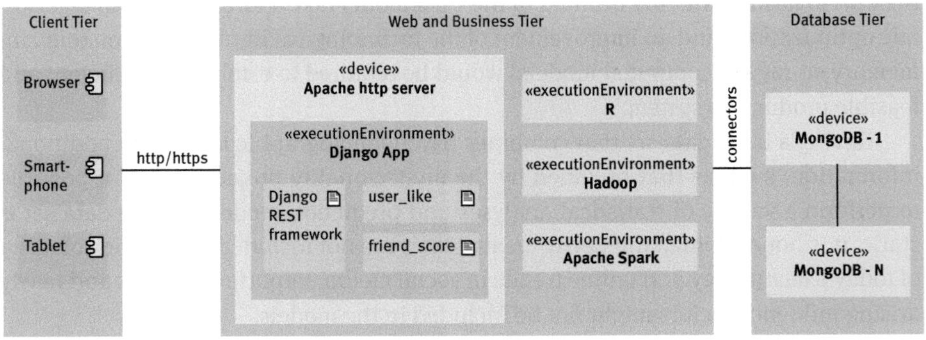

Fig. 1.8: Potential future deployment model of the next prototype

Evaluative suggestions. The proposed architecture is an initial design that is part of an ongoing search process and represents a proof of concept. The next iteration might result in another proof of concept, but a second iteration of the infrastructure could easily include software to measure the computing time, as well as the efficiency and effectiveness of the algorithms. MongoDB, for example, has a guide for query optimization that can lead to a reduction in computing time [23]. The efficiency of an algorithm can be measured during computation. Additionally, the effectiveness of an algorithm, especially one built on inferential statistics, can be measured with a set of standard tools including mean, distribution and deviation. Many other possibilities result from the potential of using already used data sets and trying different algorithmic models on them, to see which ones optimally fit the attainment of a predefined goal. Ineffective algorithms could be compromised by the unsuitability of the underlying methods, but also by bias in the data. No matter what the output might be, it is clearly preferable to measure each modification so that the infrastructure and the algorithms can be optimized.

1.5.3 Concluding remarks

Big data are a mirror of personal identity. It might seem mystical, sometimes infeasible, and for most people something inconveniently hidden in technologies they only superficially understand. Constantly assessed, data are unavoidably becoming a core asset of any organization or public authority. The goal of our chapter was to introduce the reader to a technological demonstration that data-driven economies and governments have the ability to collect private information about citizens and to build psychometric analytical models that predict human behaviour. These analyses enable, for example, political parties to send targeted messages to individuals that vary according to each individual's characteristics. This might result in the manipulation of the wills of citizens. So far, communities have lacked judiciousness, adequate jurisprudence and law enforcement structures to take effective countermeasures against threats that result from modern technological innovations, such as psychometric micro-targeting.

Bibliography

[1] McAfee A and Brynjolfsson E. *Machine, platform, crowd: Harnessing our digital future*. W. W. Norton & Company, New York, first edition edition, 2017.

[2] Fosso Wamba S, Akter S, Edwards A, Chopin G, and Gnanzou D. How 'big data' can make big impact: Findings from a systematic review and a longitudinal case study. *International Journal of Production Economics*, 165:234–246, 2015.

[3] Kranish M. Trump's plan for a comeback includes building a 'psychographic' profile of every voter; https://www.washingtonpost.com/politics/trumps-plan-for-a-comeback-includes-building-a-psychographic-profile-of-every-voter/2016/10/27/ (accessed 19.12.2017), 2016.

[4] Rust J and Golombok S. *Modern psychometrics: The science of psychological assessment*. Routledge, Hove, East Sussex and New York, 3rd ed. edition, 2009.

[5] Gerber AS, Huber GA, Doherty D, and Dowling CM. The big five personality traits in the political arena. *Annual Review of Political Science*, 14(1):265–287, 2011.

[6] Goldberg LR. An alternative "description of personality": The big-five factor structure. *Journal of Personality and Social Psychology*, 59(6):1216–1229, 1990.

[7] Wong JC, Levin S, and Solon O. Bursting the facebook bubble: we asked voters on the left and right to swap feeds; https://www.theguardian.com/us-news/2016/nov/16/facebook-bias-bubble-us-election-conservative-liberal-news-feed (accessed 19.12.2017), 2016.

[8] Cohen RJ, Swerdlik ME, and Sturman ED. *Psychological testing and assessment: An introduction to tests and measurement*. McGraw Hill, New York, 8th ed., international student ed. edition, 2013.

[9] Eysenck HJ. *The structure of human personality*. Methuen, New York, 1953.

[10] Gerber AS, Huber GA, Doherty D, Dowling CM, and Panagopoulos C. Big five personality traits and responses to persuasive appeals: Results from voter turnout experiments. *Political Behavior*, 35(4):687–728, 2013.

[11] Vecchione M, Schoen H, Castro JLG, Cieciuch J, Pavlopoulos V, and Caprara GV. Personality correlates of party preference: The big five in five big european countries. *Personality and Individual Differences*, 51(6):737–742, 2011.

[12] Dennison J. Populist personalities? the big five personality traits and party choice in the 2015 uk general election; http://blogs.lse.ac.uk/politicsandpolicy/populist-personalities-the-big-five-personality-traits-and-party-choice-in-the-2015-uk-general-election/ (accessed 19.12.2017), 2015.
[13] Blei DM, Ng AY, Jordan MI, and Lafferty J. Latent dirichlet allocation. *Journal of Machine Learning Research*, 3(1):993–1022, 2003.
[14] Blei DM. Probabilistic topic models. *Communications of the ACM*, 55(4):77, 2012.
[15] Kosinski M, Wang Y, Lakkaraju H, and Leskovec J. Mining big data to extract patterns and predict real-life outcomes. *Psychological methods*, 21(4):493, 2016.
[16] myPersonality; http://mypersonality.org/wiki/doku.php (accessed 19.12.2017), 2016.
[17] Underwood T. The stone and the shell; https://tedunderwood.com/2012/04/07/topic-modeling-made-just-simple-enough/ (accessed 19.12.2017), 2012.
[18] Pennebaker Conglomerates I. Linguistic inquiry and word count (liwc); https://liwc.wpengine.com/ (accessed 19.12.2017), 2017.
[19] Farnadi G, Zoghbi S, Moens MF, and De Cock M. Recognising personality traits using facebook status updates. In *Proceedings of the workshop on computational personality recognition (WCPR13) at the 7th international AAAI conference on weblogs and social media (ICWSM13)*. AAAI, 2013.
[20] Long B and Chang Y. *Relevance ranking for vertical search engines*. Newnes, Waltham, Massachusetts, USA, 2014.
[21] Park G, Schwartz HA, Eichstaedt JC, Kern ML, Kosinski M, Stillwell DJ, Ungar LH, and Seligman MEP. Automatic personality assessment through social media language. *Journal of personality and social psychology*, 108(6):934, 2015.
[22] Wallach HM, Mimno DM, and McCallum A. Rethinking lda: Why priors matter. In *Advances in neural information processing systems*, pages 1973–1981, 2009.
[23] MongoDB, inc.; https://www.mongodb.com/nosql-explained (accessed 19.12.2017), 2017.
[24] Dean J and Ghemawat S. Mapreduce: simplified data processing on large clusters. *Communications of the ACM*, 51(1):107–113, 2008.

Hrishikesh Bhaumik*, Siddhartha Bhattacharyya, Surangam Sur, and Susanta Chakraborty

2 Keyframe selection for video indexing using an approximate minimal spanning tree

Abstract: Video summarization is an important field of research in content-based video retrieval. One of the major objectives in this domain has been to generate the summary of a given video in the shortest possible time. In this work, the primary aim is to rapidly select keyframes from the component shots of a video to generate a storyboard in minimal time. The time taken to produce the storyboard is directly proportional to the number of correlations to be computed between the frames of a shot. To reduce the time, user input is obtained regarding the amount of actual correlations to be computed. The other correlation values are predicted by interpolation from the actual correlation values computed for the purpose. Keyframes are selected from each shot by generating an approximate minimal spanning tree (AMST) and computing the density around each frame of the shot by means of an automatic threshold based on the statistical distribution of the correlation values. From the set of keyframes obtained, redundant frames are eliminated using the SURF feature descriptor to arrive at the final summary, which depicts the storyboard of the video. A comparison between the summaries produced by an AMST and MST show encouraging results and demonstrate the usefulness of the proposed method.

Keywords: Storyboard generation, Approximate minimal spanning tree, Interpolation, Redundancy elimination

2.1 Introduction

Content-based retrieval systems (CBRS) have been the focal point of research for the past two decades. CBRS is a broad research field that encompasses various other domains like content-based video retrieval systems (CBVRS) [1], content-based image retrieval systems (CBIRS) [2] and content-based audio retrieval systems (CBARS) [3]. 'Content-based' refers to the semantic content of the media type under consideration and 'retrieval' refers to the act of extracting the relevant data (audio, video, or image) required by the user. Out of all the types of CBRS, CBVRS is the most challenging one

***Corresponding author: Hrishikesh Bhaumik,** RCC Institute of Information Technology, Kolkata, India
Siddhartha Bhattacharyya, RCC Institute of Information Technology, Kolkata, India
Surangam Sur, Jadavpur University, Kolkata, India
Susanta Chakraborty, Indian Institute of Engineering Science and Technology, Shibpur, Howrah, India

https://doi.org/10.1515/9783110551433-002

because it encompasses the analysis of all other media types like text, image and audio. Even video clips embedded within a video are quite common these days. The field of CBVRSs has drawn much attention and contributions from researchers and application developers in several domains. Research in CBVRSs has been propelled by the unrestrained uploading of videos on several online repositories, augmented by the ever-decreasing cost of storage devices.

In the early stages, CBVRS development occurred in the form of academic research projects such as the Informedia Digital Video Library from Carnegie Mellon University and Fischlar Digital Video Suite from Dublin City University. These systems operated over thousands of hours of video repositories. The important aspects for ensuring the efficacy of such systems are as follows:
1. Acquiring a concise and complete representation of video sequences
2. Providing the user with different search techniques depending upon the type of search required

The essence of research in this field is to develop different search techniques for efficient mining of videos from a huge digital repository. The extraction of relevant items from such a digital library is based on search by semantic content, rather than metadata tagged to the videos. The mining process is facilitated by the extraction of features obtained from the component units of such video data. The component elements considered for CBVR activity may consist of scenes, shots or frames of the video under consideration. Temporal decomposition at different granularity levels gives rise to one of the three components of a video, i.e., scenes, shots or frames, as depicted in Figure 2.1. While frames represent the lowest granularity level, shots are a collection of time-sequenced frames that are part of an uninterrupted sequence taken by a single camera. At the highest level of video decomposition, scenes represent a conglomeration of a number of temporally consecutive shots that are semantically similar. Semantic similarity arises when different shots share the same background or objects composing a scene.

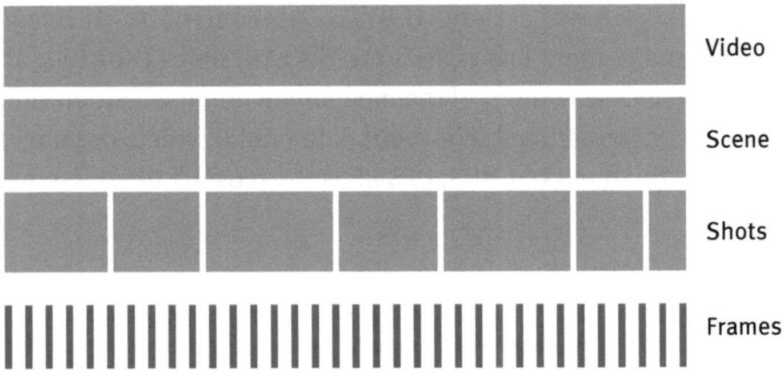

Fig. 2.1: Structural hierarchy of a video

It is imperative that the videos stored in large digital repositories be indexed in order to enable the efficient retrieval of relevant items from the database. An indexed digital library allows easy searching, retrieval and browsing of relevant videos in accordance with user requirements. The process adds flexibility and ensures that the unstructured multimedia content is both accessible and searchable [4]. The indexing process may start as soon as a video is stored in the digital library. Video indexing may be done using any of the three component units. The type of features required for indexing videos grows in complexity (low level to high level) as the level of granularity decreases. Low-level features in the form of pixel statistics like mean, variance, standard deviation of values [5], histogram of pixel values [6], entropy [7], luminance, fuzzy hostility index of pixels [8] and so forth may be extracted from high-granularity units (frames). High-level features such as optical flow in successive images [9], shape of video objects [10], edges in objects [11], motion vector [12], event detection [13], action modelling [14] and so forth are more appropriate for representing the semantic content of a shot or scene. This includes the identification or tracking of objects [15], action recognition [16], emotion detection [17] and so forth.

Video indexing using frames gives rise to a static representation of contents, which is referred to as static video summarization or storyboard. In a storyboard representation, a subset of frames from the entire set of component frames in a video is chosen to portray the contents. The process of selecting the keyframes is based on algorithms formulated for the purpose. The two aspects considered in a summarization activity are conciseness and coverage. These two aspects are in conflict with each other since conciseness affects coverage and vice versa. Indexing videos in a repository using shots or scenes allows for a dynamic representation, also known as video skimming. In such a representation, a few shots are chosen from the entire set of shots that make up the video. However, any visual abstract of a video must satisfy the following three postulates:

1. Video summary must hold the higher entities and events of interest
2. It should present rational or logical points of flow or progression
3. It should be free of recapitulation (redundancy-free summarization)

The present approach deals with keyframe selection from videos of a large repository in order to facilitate easy indexing, retrieval and browsing. Over the years, keyframe selection has been a frequently used technique for video indexing. Efficient selection of keyframes from a video helps to portray the contents in a summarized form (also referred to as static summary) to the user. This form of video summarization is particularly useful in bandwidth-constrained scenarios where the user needs to retrieve relevant videos in the minimum possible amount of time.

In the present work, an investigation is carried out to improve the performance of the method based on minimal spanning trees (MSTs), developed by Bhaumik et al. in [18]. The main motivation of this work is to reduce the time complexity for generating MSTs pertaining to shots having a large number of frames. It is common knowledge that the time complexity of MSTs is $O(n^2)$. The focus of this work is to develop a method for generating an approximate MST (AMST) based on interpolation. Given a subset of actual distances, the remaining distances for generating an AMST are predicted using interpolation. The number of actual distances computed for the purpose is based on an input percentage specified by the user. It is obvious that a high value of input percentage will contribute to reducing the error induced in the AMST while increasing the time for generating the AMST. An input percentage of 100 will generate an AMST that would be identical to the original MST and the error induced would be zero. The objective here is to generate the summaries obtained using both MST and AMST and compare it with the user-generated summaries, in terms of well-known metrics such as precision and recall.

The remainder of this chapter is organized as follows. A survey of related works on static and dynamic video summarizations is presented in Section 2.2. Basic concepts and definitions are given in Section 2.3. The proposed methodology for static video summarization, along with an explanation of the different steps of the algorithm, is enumerated in Section 2.4. The experimental results of the summarization process using the proposed methodology are elaborated in Section 2.5. Finally, some conclusions and remarks are given in Section 2.6.

2.2 Survey of related works

Video summarization is one of the most important approaches to speedy browsing and accessing of videos from large digital repositories. It is the process of constructing a short summary from a large digital video to give users a useful visual abstract about the contents of the video. The component units of a video can be defined as scenes, shots and frames depending upon the granularity level used. Thus, the temporal decomposition of a video gives rise to different coarseness levels corresponding to the stages of video hierarchy. As shown in Figure 2.1, several units at a lower stage are combined to form a unit at a higher stage. The decision to combine a set of units at a lower level of the hierarchy to form a unit of the higher level depends upon the similarity of such units, given the features extracted from each unit. This results in the different domains of video analysis such as shot boundary detection or shot transition detection, scene detection and so forth. The sequential ordering of the temporally related units in a video is of prime importance because any disordering will cause a deformation of the content [19]. Video summarization may be classified into two broad categories: static video summarization and dynamic video summarization.

2.2.1 Static video summarization

The term 'static' literally means stationary, stable or fixed. The process of static video summarization involves the selection of a certain number of frames from a video. The process comprises the following basic steps:
1. Decompose video into images using a codec.
2. Extract features from each frame composing the video.
3. Select a set of representative frames based on features extracted from each frame.
4. Reduce redundancy on set of representative frames to eliminate superfluous frames.

The static summary reflects the basic underlying content of a video. The summary generated may be a set of representative frames or a collage formed from such frames. In the literature, static video summarization has also been referred to as storyboard generation, which has attracted the attention of many researchers, who identified a need to represent the contents of a given video in concise form. The two basic approaches used in video summarization and detailed in later subsections are:
1. Classical approaches
2. Soft computing–based approaches

2.2.1.1 Classic approaches

Clustering has been used by many researchers as an effective tool for generating video summaries. Mundur et al. [20] applied the concept of Denaulay clustering to produce effective static summaries of videos. In this method, Denaulay triangulation is used to cluster frames, which produces good-quality summaries with few frames and reduced redundancy. In this method no user intervention is required, and the method is well suited for batch processing. In [21], De Avila et al. focused on the extraction of colour features from video frames for clustering the same using a k-means clustering algorithm. This algorithm was called VSUMM and the generated summary was compared with a manually generated summary by human users. The comparison revealed that the system had produced a high-quality summary. In [22], a static video summarization algorithm was presented that was based on a k-medoid clustering method. The distance measure taken for the purpose is computed on a block matching mechanism based on frequency domain. The approach is computationally viable and is able to extract dissimilar frames in large motion scenarios. The keyframes extracted by the process have reduced visual redundancy. In [23], Hanjalic proposed a scheme for producing a static and dynamic summary by applying multiple partitional clustering to all frames of a video. After cluster validity analysis, centroids from the optimal clusters thus obtained form a set of representative frames or storyboard. The shots corresponding to the selected keyframes are coalesced together to form a dynamic summary.

Cahuina and Chavez proposed a method [24] for video summarization taking into consideration local descriptors and temporal video segmentation to produce static video summaries. The evaluation was carried out on more than 100 videos, and the approach yielded good results compared to other state-of-the-art methods. The experiments conducted also conveyed the importance of colour information. Bhaumik et al. in a paper [25] introduced a three-phase keyframe extraction approach to generate a static summary of a video. It is one of the few works that focus on redundancy reduction of detected keyframes. The algorithm works in three stages, where the first stage focuses on the selection of representative frames and the other two stages focus on the intrashot and intershot redundancy reduction of the keyframes selected in the first stage. After every stage, a comparison of the system generated and user-generated summaries is done. In [26], the authors take a set-theoretic approach to the problem of keyframe extraction from a video. The extraction algorithms proposed in the work enable the presentation of a summary in the form of a multilevel abstract. For a given degree of fidelity, a tree-structured hierarchy is generated that enables well-organized content retrieval using a depth-first search. The work by DeMenthon et al. [27] on generating representative frames of a video focuses on a curve simplification method. The idea behind this work is to represent a video as a trajectory in high-dimensional feature space, where each of the dimensions represents a particular feature of the video frame. The trajectory represents a temporal variation of the features in the feature space. To select the representative frames, the curve is broken down into several component arcs. The frames at the junction of the arcs are selected as representative frames. The approach is versatile in that the level of decomposition can be controlled, thereby decreasing or increasing the number of keyframes for summarizing the video.

The work by Yeung and Yao [28] is one of the early papers on video summarization that speaks about the objectives of the summarization task and presents a technique for computing the dominance of contents in various segments of a video and thereby generating a pictorial summary from the chosen video frames. Franti et al. introduced two methods based on a Hough transform for image matching in [29]. In the first one, a feature vector is formed by adding the most important coefficients in each column of the accumulator matrix. Thus, this method represents an easy approach to matching that is invariant to geometric transformation. The second approach takes into account the positional data of lines to provide a more accurate depiction of images. Thus, this method provides an effective way to match two images, which is a fundamental requirement for video indexing systems. In a different approach, a Gabor filter was used by Zhang et al. [2] for extracting the textural features of an image. Rotational normalization is accomplished by circular shifting, so that the images have the same dominant direction. With this approach, images can be indexed for retrieval of similar images from a large digital repository. Gygli et al. [30] used a supervised learning approach for the characterization of a summary that optimized multiple objectives for producing a good summary. In another work, Xu et al. [31] adopted gaze tracking information for summarizing personal videos. Itazuri et al. [32] proposed a video summarization scheme for volleyball videos using a rally scene detection method based on court changeover information. A multimodal video database and movie summarization system was introduced by Zlatintsi et al. [33] for the evaluation of video summarization algorithms. Graph modularity clustering was used in [34] for the extraction of keyframes. Sharghi et al. [35] focused on the aspect of user subjectivity to create a video summary. In this approach, a memory network parameterized sequential determinantal point process was used to address user queries on the frames and shots of a video. A two-stage model was proposed by Lin et al. [36] for the online summarization of videos for the Internet of Video Things (IoVT), captured at the sensor nodes, in which a two-layer *k*-nearest neighbour clustering algorithm is used for object clustering. A keyframe recommender system was designed by Chen et al. [37] that could model both textual and visual features in a unified framework. In this approach, the personalization provided to the user was achieved by encoding several user interests into a unified multimodal space. Sahoo et al. [38] devised a video summarization system that could produce static, dynamic or entity summaries. The system is able to summarize videos on the fly. On-the-fly summarization of large videos was taken up in the work by Mirzasoleiman et al. [39] using a single-pass streaming algorithm designed for the purpose. In the approach by Choi et al. [40], context-driven video summaries are generated by learning the semantic embeddings of video frames by means of captions present in frames. The system is able to produce a meaningful summary by selecting relevant video segments, given a textual description of user needs. The system developed by Otani et al. [41] takes into account user intentions for posting a blog video by focusing on the supporting texts in the blog. The summaries generated by the system reflect the blogger's perspective. In [42], Mitra et al. used sequences of temporal segments called tracklets for discovering entities in a video. The tracklets are clustered based on parameters like clustering accuracy, cluster purity and entity coverage. The method is applicable for streaming videos and is also able to reject false alarms. The approach used by Mademlis et al. [43] focuses on summarizing human activity in videos by identifying visual building blocks and forming basis vectors. A joint optimization model is extracted from a column subset selection problem (CSSP) to provide content coverage and inclusion of outliers. In [44], a half-quadratic minimization algorithm is used to solve a collaborative sparse optimization problem for clustering large collections of videos based on the given topic. In this work, techniques from information retrieval and natural language processing are used for collaborative summarization. A technique for real-time video summarization was proposed by Abdelali et al. in [45] that used an MPEG-7 colour structure descriptor. The system was implemented on a chip using a Xilinx Virtex5 field-programmable gate array.

A detailed survey of video summarization made with the help of wearable cameras is presented in [46] by Del Molino et al. A supervised video summarization framework was developed by Ji et al. [47] in which an attentive encoder-decoder network was developed for automated video summarization. Bidirectional long short-term memory (BiLSTM) was used to encode the contextual information in input video frames. The decoder uses two attention-based LSTM networks for additive and multiplicative objective functions. Video summarization using an ensemble of deep learned features was tested in [48]. In [49], Panda et al. focus on the summarization of a collection of videos rather than only one video. The method explores the complementarities in a given set of videos by means of a separate minimization algorithm over an objective function. In a separate work [50], the authors took up the problem of summarizing surveillance videos taken from multiple cameras. The work focuses on multiview correlations that serve to represent a diverse set of representative shots. Also, sparseness is modelled to select representative shots in order to produce a summary. In [51], a fast iterative shrinkage thresholding algorithm (FISTA) was used by Meng et al. for category-specific video summarization using a multi-view sparse dictionary selection with centroid co-regularization (MSDS-CC) method. An approach for object-level video summarization was formulated by Zhang et al. in [52] by means of an online motion Auto–Encoder (online motion-AE) framework that executes on the segmented object motion snippets of the video. The encoder imitates the online dictionary learning by memorizing previous states of object motion; it updates itself through a recurrent auto-encoder network. Instead of using a single-source media type, Rudinac et al. in [53] focus on using dynamic multisource media that may be more effective at summarizing content from modern social media. Tejero-de-Pablos et al. in [54] deal with the problem of automated sports video summarization using a deep neural network to extract action-related features and to categorize video fragments into interesting and uninteresting segments. A method for video summarization was developed in [55] by Plummer et al. that uses a joint vision-language approach to evaluate the representativeness and interestingness objectives.

In the next section, video summarization approaches using soft computing are discussed.

2.2.1.2 Soft computing–based approaches

1. Approaches based on fuzzy logic

Researchers in the field of video summarization have frequently used fuzzy logic to enhance the performance of systems developed for the purpose. In an interesting work by Fang et al. [56], shot boundaries were detected using fuzzy logic in order to integrate hybrid features constructed from basic features like pixel differences, colour histogram, motion of objects and edge features. The approach uses a mode selector to detect hard cuts and gradual transitions in two different processing modes. A data set from Carleton University is used to evaluate the system. In another approach, Das et al. [57] used low-level fuzzy features to generate a fuzzy model for detecting shot transitions. The method is useful for detecting abrupt transitions as well as different types of gradual transitions like dissolves, wipes and fades. The method achieves high precision and recall values on test videos, especially for gradual transitions. Anderson et al. [58] proposed a method based on fuzzy logic for event detection. The work deals with the problem of fall detection of humans using a 3D object generated from multiple cameras. The states of a person in every image are taken, and a linguistic summarization of the person's state is given as output using a fuzzy rule base. The framework is flexible since the existing rules can be modified or removed and new rules added. In [59], Cahuina et al. presented a method for static video summarization based on colour histograms generated from a video. Principal component analysis is used for dimensionality reduction of feature vectors. Thereafter, the shots in the video are detected using the Fuzzy-ART algorithm [60]. The keyframes are extracted using the Fuzzy C-Means [61] algorithm. A multidimensional fuzzy histogram–based approach to video summarization and retrieval is discussed in [62] by Daulamis et al., in which the visual content is portrayed for a video. In this approach, suitable features are extracted to produce a multidimensional fuzzy histogram. Video summarization is achieved by eliminating shots or frames with similar content. During the retrieval process, images that visually match a query are extracted from video frames. An interesting approach to home video summarization that employed a fuzzy fusion method for combining the attention and emotional aspects of a user was adopted by Peng et al. in [63]. The system used an Interest Meter (IM) that estimated the attention aspect of a user by tracking head motion, eye movement and blinking. To estimate the emotional state of a user, facial expressions are captured and classified as positive or neutral. A quantitative interest score is generated for different parts of the video. The summarized version contains those parts of the video that have a high interest score. A video surveillance system for managing elderly and disabled people was proposed by Elbouz et al. [64]; it used fuzzy logic to detect a person's behaviour. In this system, the Vander-Lugt correlator (VLC) and joint-transfer correlator (JTC) techniques are integrated in order to make a decision about the 3D poses of a person. Since no extra devices are involved, the system is cost effective. Model-based prediction and optical correlation are used to track the person by locating the position of the person's eyes in the image. The object in the image is detected using HSV colour space. In another work with a similar motivation, aimed at monitoring elderly and disabled people, Zhou et al. [65] developed a system for activity recognition and video monitoring. The physical location and speed of a person are estimated using a learning model developed on the basis of features extracted from a video. A human detection and tracking algorithm is developed at the object level. For action recognition, a hierarchical decision tree and dimension reduction methods were used. Thus the functional assessment of a person is estimated by extracting the statistics related to activities of daily living. The approach was tested on more than 200 h of video and shown to be effective for this type of application.

Keyframe selection using a stochastic framework was presented by Avrithis et al. [66] in which feature-based representation is obtained through colour and motion segmentation on each video shot. The recursive shortest spanning tree algorithm is used in the approach. A fuzzy multidimensional histogram is used to assemble all segment features so that misclassification of segments can be decreased. The work illustrates two approaches to keyframe detection. The trajectory pertaining to the temporal variation is analysed in the first method. The second approach aimed at decreasing cross-correlations. For the second method, a stochastic adaptation of logarithmic searches and a genetic algorithm were designed. A similar kind of work was carried out by Doulamis et al. [67] in which an effective algorithm for video summarization was developed. The approach took into consideration the temporal variations of video content. Important frames or shots were extracted by approximating suitable points on the feature vector curve and the curve formed by interpolation using the estimated points.

2. Approaches based on neural networks

Neural networks have been effectively used by researchers for the summarization of video content. Convolution neural networks (CNNs) have been found to be particularly useful in dealing with image recognition problems. Karpathy et al. applied CNNs to the problem of large-scale video classification in [68]. To take advantage of local spatio-temporal information, multiple approaches are used to extend the connectivity of a CNN in the time domain. A large data set consisting of one million YouTube videos having 487 class labels were used for experimentation, and a significant improvement in performance was achieved. Recurrent neural networks (RNNs) have also been used effectively for image recognition problems. Yao et al. [69] used local and global temporal structures to produce a description of videos. Short temporal dynamics are modelled with a 3D CNN so that it can be used to train the action recognition tasks in line with human actions. Also, a text-generating RNN is used to produce a natural language description of video segments using a temporal attention mechanism. Venugopalan et al. have also used CNNs and RNNs in the description of videos in natural language in [70]. The approach uses deep neural networks to create natural language descriptions of open-domain videos with large vocabularies. Training for the neural networks is carried on a large data set of images having category labels as well as other images bearing captions.

An adjustable neural network model was proposed by Doulamis et al. in [71] for video object (VO) segmentation. In this approach an adaptive neural network classifier is used for object tracking. This is seen to be more effective than conventional object motion tracking algorithms. A cost-efficient algorithm is used for weight updating of networks without degrading the knowledge acquired by the network. The detection of the body and face of human objects is attained by Gaussian distributions for video conferencing applications. To detect generic VOs, colour and depth information is extracted for segmentation fusion. Effective results are obtained for scenarios where there is object occlusion or blending.

A six-phase method based on machine learning was proposed by Zawbaa et al. [72] for event detection and summarization of football (American soccer) matches. In this approach, both support vector machines (SVMs) and neural networks are used for the detection of important video segments. The system performs scoreboard detection, replay detection, exciting event detection and logo detection for the summarization of football videos.

In [73], Panda et al. modelled a 3D CNN that is able to automatically learn a parametric model for video classification. The model is used to detect important segments of a video in order to produce a summary. RNNs were used by Sah et al. [74] to generate textual summaries using deep learning models. The significant portions of videos are converted to textual summaries using recurrent networks. In [75], Sanghavi and Pangaonkar adopt a self-organizing map (SOM) to detect indoor and outdoor scenes for efficient video summarization.

3. Approaches based on genetic algorithms

The need to summarize large videos has posed a long-standing challenge in the field of video indexing and retrieval. Most traditional methods fail on account of issues related to time complexity. In the analysis of large videos, fast browsing and access to meaningful content are integral parts of the research in the field. In this respect, Chiu et al. [76] devised a genetic segmentation algorithm that executes iteratively on the different segments of representative strings. The method facilitates segmentation in an incremental manner due to the evolutionary nature of genetic algorithms. This enables summarization of long videos in an efficient manner for faster indexing, browsing and retrieval. The effectiveness of content-based retrieval systems can be enhanced by combining multiple features extracted from images. However, the combination of features may not always prove effective, and the combined function may be far more complex than simple weight-based functions. To address this problem, Torres et al. [77] proposed an effective way to achieve a non-linear combination of image features in a genetic programming framework. In this method, images are retrieved based on object shapes. Lu et al. in [78] presented an approach in which image feature descriptors for colour, texture and shape are selected as low-level features of the image. Three types of chromosomes, i.e., real coded, binary and bi-coded chromosomes, along with k-nearest neighbour classification accuracy, are used as fitness functions to optimize the weights and selection of descriptor subsets. A data set of 2000 annotated Corel images was created for experimentation, and it was found that the annotation accuracy increased by 7%, 9% and 13.6% for the real coded, binary and bi-coded chromosomes respectively.

A scene retrieval algorithm was proposed by Yoo and Cho [79] in which five video features, i.e., average colour histogram, average brightness, average edge histogram, average shot duration and gradual change rate, were extracted from videos and mapped to the user's emotional space using an interactive genetic algorithm. The emotion categories considered here are action, excitement, suspense, quietness, relaxation and happiness. An initial population is selected by the algorithm based on the emotions given by the user. The feature vectors from this initial set are treated as chromosomes, and genetic crossover is applied to produce a new chromosome set. Using a similarity function, videos corresponding to the new chromosomes are retrieved. This process is repeated until the videos that are relevant to the user are retrieved. A method for selecting keyframes based on CSSP was formulated by Mademlis et al. [80]. In this case, the problem is viewed as an optimization problem and solved by means of a genetic algorithm.

4. Support vector machines

SVMs have been an important tool for researchers in most machine learning approaches. SVMs can serve to memorize knowledge gathered from past facts and use it for future computation. SVMs have been found to be effective for the classification of both linear and non-linear data. In this context, an interesting approach was adopted by Ng et al. [81] for modelling the relationship between brainwaves and audio/video features using an SVM model. In this approach, summarization of personal videos, taken with wearable video cameras, was done using low-level video and audio features. A different approach to video summarization was adopted by Li et al. [82] in which a mapping of video features to their corresponding textual information was attempted. The learning mechanism used transferable structure learning using a structural SVM as a backbone to enhance the summarization of videos and boost performance by exploiting the interchange between video features and equivalent textual information. To summarize a new video, only video features need to be extracted.

Potapov et al. stressed the need for category-specific video summarization in [83]. An SVM classifier was used to allocate scores to the different video fragments obtained after video segmentation. The segments having the highest scores are coalesced to produce a video summary. Experiments with TRECVID 2011 data show the efficacy of the proposed mechanism. Video segmentation using SVMs was tested in [84], which describes a fast method for shot boundary detection. Initially, the grey variance is extracted from video frames. The frames at the sequences of smooth variations are deleted and a new ordering of the frames, called a reordered frame sequence (RFS), is carried out. Pattern vectors are formed from three features, that is, pixel-wise intensity differences, colour histogram differences in HSV space and edge histogram differences in the X and Y directions are extracted from RFS and given as input to the SVM to determine the shot boundaries. Gradual transitions are detected using temporal multi-resolution on the frames of RFSs.

Producing a synopsis of sports videos is a specific and important research field in the domain of video summarization. In [85], a mechanism for detecting notable events in sports videos was devised by Sadlier and O'Connor that was based on audio-visual features. The selected features are characteristic of any sports video and are merged with the help of an SVM that serves to build a model during the training stage for the detection of important events in a given sports clip.

To overcome the problem of relevance feedback systems, where obtaining feedback from the user might be a problem due to time constraints, a method for pseudo-labelling of images was devised by Wu et al. [86] that was based on a fuzzy version of SVMs. Learning from a few training samples prevents the system from performing well. To overcome this problem, the images were psuedo-labelled. However, this induces uncertainty in the class information. Fuzzy SVM (FSVM) is thus used to address this uncertainty through active learning using a combined model called pseudo-label fuzzy support vector machine (PLFSVM) to perform content-based image retrieval. This method has proved to be effective on a large data set consisting of 10,000 images.

2.2.2 Dynamic video summarization

Due to the massive volume of video data available online, it is becoming increasingly important to devise tools and applications for summarizing video content. The need for video summarization is obvious from the fact that browsing and retrieval time must be minimized. However, in many situations a static representation of a video using keyframes may not be sufficient. In certain cases related to the summarization of sports, news and other videos, a better representation can be obtained if important shots are concatenated to produce a summary, leading to dynamic video summarization or video skimming. Techniques for video skimming encompass the following approaches.

1. Graph modelling

Video skimming was achieved through graph modelling in [87–89]. In [87], scene modelling was carried out using a normalized cut algorithm and temporal graph modelling, whereas highlight detection was achieved by motion attention modelling. In this approach, the video is represented as a complete undirected graph, and video clusters are formed by splitting the graph using a normalized cut algorithm, resulting in the formation of directed temporal graphs. Detection of video scenes is accomplished using a shortest path algorithm. In a temporal graph, attention values are computed and assigned to all video structures i.e., scenes, clusters, shots and subshots. This helps to generate video summaries that rely on both content and perceptual value. In an earlier work, Ngo et al. [88] used temporal graph modelling for the detection of subshots to create a dynamic video summary. To evaluate the perceptual quality of shots and clusters, a motion attention model was used.

In another approach, Lu et al. [89] used graph optimization for dynamic video summarization. The method emphasizes visual coherence and content coverage. In this method, a dynamic programming algorithm for video skimming using spatio-temporal information to determine the key shots and scenes in a video was proposed. The method comprised three stages. In the first stage, the given video is fragmented into shots and a candidate shot is chosen. The spatio-temporal association between the shots is modelled as a directional graph. Finally, the longest path in the graph, determined by an algorithm based on dynamic programming, is used to represent the final video summary.

2. Motion modelling

Motion modelling is an alternative method for dynamic video skimming. In [90], video segmentation is achieved by identifying dominant motion changes in the frames of a video. A 2D affine model is generated to examine some of the coefficient changes with respect to time. The residual motion content of the obtained temporal units is made a statistical representation. A set of labelled classes is available through off"|line learning from a training set of video clips. Maximum likelihood is used for the classification of the video segments. A visual motion modelling approach was also used in [91] to identify dynamic video contents for dominant image motions as well as residual image motions. An affine motion model and low-level motion model were employed to quantify the motions in the video. It has produce good results for both low-level and high-level motion videos, especially sports videos.

3. User attention modelling

Ma et al. [92] proposed an alternate approach to video semantic analysis, in the form of an attention-based approach. Instead of understanding video content through semantic analysis, the method focuses on modelling the viewer's attention. The generic framework presented in the work eliminates the need for complex rules based on heuristics and takes advantage of the attention models framed for the purpose.

4. Singular value decomposition

Gong et al. in [93] employed a technique for both static and dynamic video summarization that was based on singular value decomposition (SVD). A feature frame matrix was constructed to perform SVD in order to select the best discriminating features. The frames are clustered based on the reduced feature space, and a metric is devised to determine the amount of visual content in each cluster. The best possible set of keyframes is chosen from the clusters and a summarized video, based on length given by the user, is generated. In this method, the emphasis is on the visual content of the clusters, ensuring that redundancy is eliminated from the generated summary.

5. Expectation maximization

Orriols and Binefa used an approach based on an expectation maximization (EM) algorithm [94] for dynamic video summarization. Appearance-based frameworks, such as an appearance representation model, an appearance temporal evolution model and a maximum-likelihood estimation model, were applied to compute the time symmetry. A set of semantically meaningful frames was extracted to devise a generative model capable of representing sequence progressions in the video. The probabilistic approach used for this problem took into account the perceptual similarity taken from time progression models and learned appearance.

6. Relevance score

A method for selecting representative key shots or video clips from a full video for dynamic video summarization is discussed in [95]. Key shots with high concernment scores are recognized to summarize event-driven web videos by localizing the connected tags with shots from each video. The relevancy of the shots is determined according to the event query by connecting localized tags with the query. The method is tested on a data set consisting of more than 10,000 web videos from 60 classes.

2.3 Basic concepts

2.3.1 Shot boundary detection

A shot is a series of temporally separated sequences of frames taken by a single camera. In video clips, a shot is therefore a manifestation of continuous action without a break in sequence. The consecutive frames within a shot bear a high correlation. A change in sequence is denoted by a transition from one shot to another. The type of transition is determined at the editing phase during the production of a video from clips during the shooting of the sequences. During video analysis, these shots must be separated again by determining the shot boundaries. Shot boundaries are mainly classified into two broad categories, gradual transitions and abrupt transitions [4]. Shot transition detection is the most basic step for any video analysis application and plays a major part in video processing [96]. The main objective of shot detection is to automatically catch the transition point between two shots within a video. Detection of gradual transitions is a more challenging task than finding abrupt transitions. A lot of research has been done in this field, and several methods for the detection of both abrupt [97–99] and gradual transitions [100–103] have been developed by researchers.

An abrupt shot change or hard cut [104–106] occurs when there is an instantaneous transition from one shot to another. Abrupt shot changes account for 95% of all shot transitions [107]. In the case of abrupt transitions, the last frame of the previous shot and the first frame of the next shot are placed sequentially for amalgamation in order to achieve this type of transition.

Gradual shot change [108] occurs slowly over a few frames as one shot ends and another begins. The frames involved in a gradual transition are a combination of the same number of frames taken from both shots. The manner of combination of the frames determines the type of gradual transition, i.e., dissolve, fade-in, fade-out or a wipe effect, as portrayed in Figure 2.2. In this type of transition, frames from two shots may be combined using chromatic, spatial or spatio-chromatic effects, which causes a slow transition from one shot to another. A dissolve transition is a type of gradual shot change that occurs between two shots. There is no clearly demarcated boundary between shots. Therefore, in this type of transition, the first shot fades out and the second shot fades in, such that frames that form a part of the dissolve sequence are a linear combination of frames taken from the two shots. As a dissolve sequence progresses, the amount of contribution from pixels of one shot decreases, while the contribution of pixel values from the frames of another shot increases. Hence, a dissolve sequence can be modelled as a combination of fade-in and fade-out [109]. During a dissolve sequence, an overlapping of two or more different scenes occurs. Hence, detecting a dissolve within a video has always been a major challenge for researchers. Another category of gradual transition is the fade. Fade-in [110, 111] occurs when a mono-colour screen gives way to a scene such that the image shown on the screen progressively clears out. On the other hand, fade-outs occur when the scene being shown gradually brightens or darkens, giving way to a totally black or white screen. A wipe is another type of gradual shot change in which a line moves across the screen and a new scene appears [98]. There are a few specific types of wipes such as the iris slow, star wipe, heart wipe, matrix wipe, clock wipe and others.

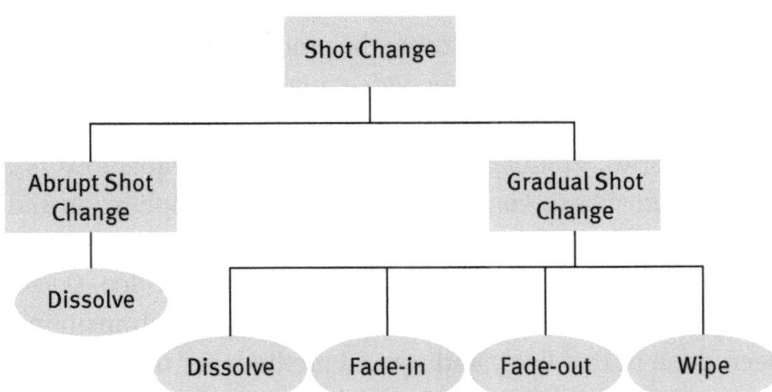

Fig. 2.2: Types of shot transitions in a video

2.3.2 Correlation between image frames

The amount of similarity between a pair of frames can be measured by comparing the corresponding pixel values of the images. The concept of image correlation was derived from wave correlations between two electronic signals. A popular method of finding correlations is by Pearson's method in which a deviation from the mean value of pixels is taken into consideration. A preprocessing stage involves the conversion of the 3D matrix of values corresponding to the RGB plane [112] to a 2D matrix containing the intensity values computed from the RGB values of each pixel. The equation for calculating the intensity value (I) is as follows:

$$I = \alpha R + \beta G + \gamma R \tag{2.1}$$

In Equation (2.1), $\alpha = 0.2126$, $\beta = 0.7152$ and $\gamma = 0.0722$. Here, I denotes a 2D matrix of values representing a greyscale image obtained from a 3D colour image. Pearson's correlation coefficient (ρ_{xy}) or a product-moment correlation coefficient is a real number whose value lies in the interval $[-1, 1]$. For two given images represented by the matrices X and Y, the amount of similarity is computed using the formula

$$\rho_{xy} = \frac{\Sigma(x_{ij} - \bar{x})(y_{ij} - \bar{y})}{\sqrt{\{\Sigma(x_{ij} - \bar{x})^2\}\{(y_{ij} - \bar{y})^2\}}} \tag{2.2}$$

Here x_{ij} and y_{ij} are the elements in the i-th row and j-th column of the matrices X and Y respectively, $\bar{x}$ is the mean value of the elements in X, $\bar{y}$ is the mean value of elements of Y and n is the total number of elements in the matrix under consideration. The standard deviation of elements in both matrices should be finite and non-zero for the correlation to be defined. The correlation is 1 if there is a perfect match of the values and -1 in the case of a decreasing linear relationship. In other words, the value of ρ denotes the degree of match between matrices [8].

2.3.3 Three-sigma rule

The standard deviation (σ) of a data set or probability distribution denotes the variation or deviation from the arithmetic mean (M) or expected value. The three-sigma rule in statistics is used to signify the range in which the values of a normal distribution are expected to be present. Figure 2.3 illustrates that 68.2% of values in a normal distribution lie in the range $[M + \sigma, M - \sigma]$, 95.4% of values in $[M + 2\sigma, M - 2\sigma]$ and 99.6% in the range $[M + 3\sigma, M - 3\sigma]$. Thus, this empirical rule may be used to establish a threshold for determining the similarity between two images. In this work, two images are considered to be visually similar if the distance (determined from the correlation) is less than $M + 3\sigma$, where M is the mean distance and σ is the standard deviation, computed from all the distances in the AMST.

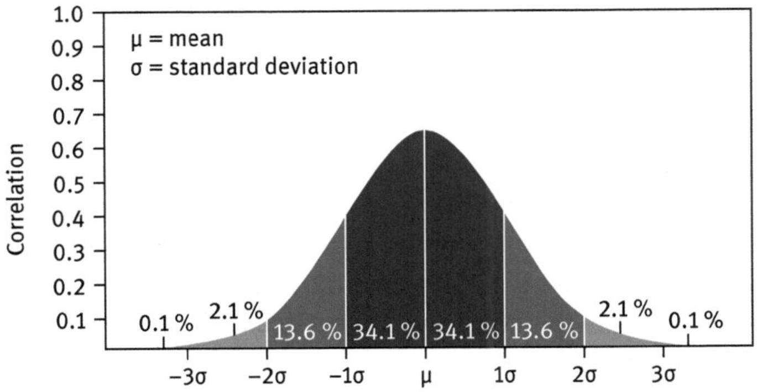

Fig. 2.3: Three-sigma rule

2.3.4 Features considered for identifying keyframes

Keyframes are a subset of the frames composing a video. These frames carry the bulk information that is crucial for summarizing a video. The keyframe set of a video is composed of reference points on the basis of which a video can be indexed. In multimedia production, a keyframe or a group of keyframes holds special information that defines a shot and can be used to discriminate between shots in a set. Often, the intermediate frames of a shot can be interpolated over time to create the illusion of motion. The features that can be extracted from a set of frames can be classified into colour-based, texture-based and shape-based attributes.

Colour-based features may consist, for example, of colour histograms, colour moments, colour correlograms or a mixture of Gaussian models. The image may be split into *kXk* blocks to capture local colour information [4].

The texture-based features of an image convey information about the spatial arrangement of pixel intensities and colour in images. The texture is basically a set of measures that allows for the identification of the perceived visual layers in an image. Gabor wavelet filters may be utilized to extract texture-based data for a video search engine [2, 113].

Shape-based features are those that describe object shapes in an image and can extract object contours or regions. To represent the spatial dissemination of edges for searching videos in TRECVID2005, an edge histogram descriptor was applied in [113].

2.3.5 Basics of interpolation

Interpolation is a mathematical tool for estimating or predicting unknown values that exist between some known values. Thus, using interpolation, new data points may be created that lie within the range of a discrete set of known data points. Given a set of known points, represented by the set X and a corresponding set of values Y, where $Y = f(X)$, the task is to predict $y_i \in Y$ for some $x_i \in X$ such that $x_a < x_i < x_b$, where $x_a, x_b \in X$ and $y_i = f(x_i)$.

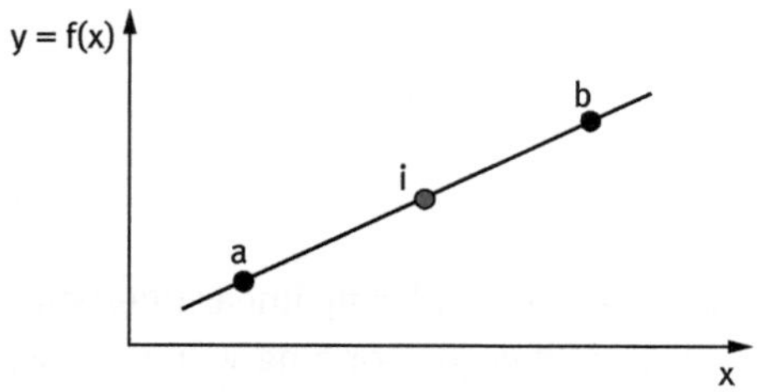

Fig. 2.4: Linear interpolation

In the graph shown in Figure 2.4, a straight line passes through two known points '*a*' and '*b*' that are involved in estimating the unknown point '*i*' that exists between these known points. The interpolated value of the point '*i*' can be estimated easily since all three points lie in a straight line. However, a typical problem that arises in interpolation is the approximation of a complicated function using a simple function. A case may arise where the formula for some given function is known but is too complex to evaluate efficiently. In such a case, the known values from the original function can be used to create an interpolation based on a simple function. However, errors may be induced if the complex function is approximated or represented by such a simple function. In reality, many types of interpolations exist, for example, linear, polynomial, spline and multivariate. In this work, linear and polynomial interpolations are examined for approximating the distances to be computed to generate an AMST.

2.3.5.1 Linear interpolation

Linear interpolation is a method of curve fitting using polynomials of degree one in order to generate new data points within the range of a discrete set of known data points. The formula for a straight line can be represented as $P(x) = a_0 + a_1 x$. A different way to characterize the same is

$$P(x) = \frac{(x - x_1)}{(x_0 - x_1)} y_0 + \frac{(x - x_0)}{(x_1 - x_0)} y_1 \tag{2.3}$$

$$= \frac{(x_1 - x) y_0 + (x - x_0) y_1}{(x_1 - x_0)} \tag{2.4}$$

$$= y_0 + \frac{x - x_0}{x_1 - x_0} [y_1 - y_0] \tag{2.5}$$

$$= y_0 + \left(\frac{y_1 - y_0}{x_1 - x_0} \right) (x - x_0) \tag{2.6}$$

To elaborate the concept of linear interpolation, the following example is considered:

Tab. 2.1: Table for tan *x*

x	1	1.3	1.5	1.7
$\tan x$	1.5574	3.6021	14.1014	-7.6966

Here, the problem is to evaluate the value of tan(1.47). Considering linear interpolation, the points of interest are $x_0 = 1.3$, $x_1 = 1.5$, with corresponding values for y_0 and y_1. Then,

$$\tan x \approx y_0 + \frac{x - x_0}{x_1 - x_0}[y_1 - y_0] \tag{2.7}$$

$$\Rightarrow \quad \tan(1.47) \approx 3.6021 + \frac{1.47 - 1.3}{1.5 - 1.3}[14.1014 - 3.6021] = 12.5265$$

Although the true value of $\tan(1.47) = 9.8874$, linear interpolation produces an error of 26.69% in the computed value.

2.3.5.2 Polynomial interpolation

Polynomial interpolation is a method that estimates the values between known data points by fitting a polynomial curve of degree n. The maximum degree of the fitted curve is dictated by the number of known points. The values of unknown points can be estimated using polynomial interpolation when these points lie between known points. A polynomial of degree n may be represented as

$$P(x) = a_n x^n + a_{n-1}x^{n-1} + \cdots + a_2 x^2 + a_1 x + a_0 \tag{2.8}$$

It can be seen from Equation (2.8) that the number of known points has to be $(n + 1)$ in order to fit a polynomial of degree n. Considering the same problem of computing tan(1.47), a polynomial of degree 3 may be fitted, resulting in the following polynomial:

$$P(x) = a_3 x^3 + a_2 x^2 + a_1 x + a_0 \tag{2.9}$$

The values of the coefficients computed by substituting the four points in Table 2.1 are computed as

$$a_3 = -707.3\,, \quad a_2 = 2778.9\,, \quad a_1 = -3562.8\,, \quad a_0 = 1492.7$$

Thus, the value of $\tan(1.47) = 13.559$ is computed using a polynomial of degree 3.

2.4 Proposed methodology

2.4.1 Flow diagram

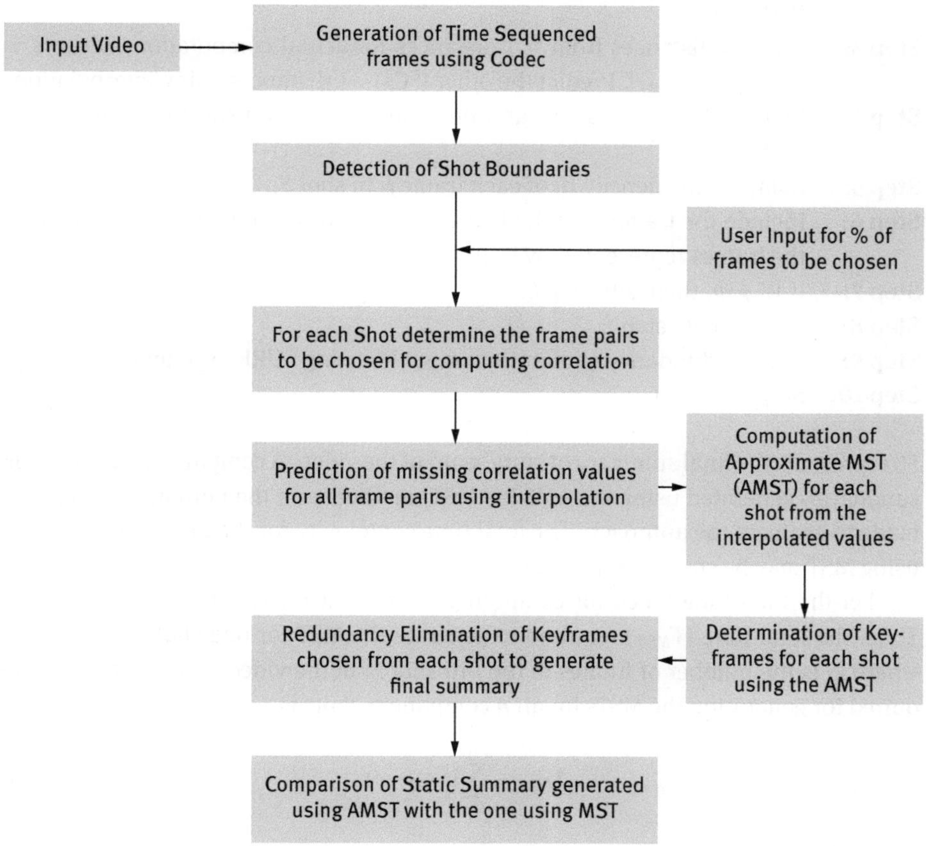

Fig. 2.5: Flow diagram for proposed method

2.4.2 Algorithm for storyboard generation

A video $V = S_1 \otimes S_2 \otimes S_3 \otimes \ldots \otimes S_{n-1} \otimes S_n$ consisting of n shots is given as input to a system. The percentage (p) of actual distances to be used is input by the user. RS represents the keyframe set generated by selecting keyframes from the component shots, and FS represents the final summary after redundancy elimination is carried out on RS. The density d_j of a frame f_j is the number of frames lying on a disc of radius

$\gamma = \mu + 3 \cdot \sigma$, which measures the representative ability of the frame. Here, μ and σ are the mean and standard deviation of the edges of the MST generated for each shot S_i.

Initialize: $i = 0$; RS $= \phi$

Step 1: Take $i = i + 1$; k = number of frames in S_i; W = set of frames in shot S_i.

Step 2: Determine total number of distances to be computed for generating MST $=^k C_2$.

Step 3: Select s distances from ${}^kC_2$ distances for actual computation, where $s = \lceil (p/100) \cdot ({}^kC_2) \rceil$. Predict the other $({}^kC_2) - s$ distances using interpolation.

Step 4: Using actual s distances and predicted $({}^kC_2) - s$ distances, generate the AMST.

Step 5: Compute the density d_j of each frame f_j in shot S_i.

Step 6: Include the frame with the highest density in W in the set RS and remove the frames represented by it from W.

Step 7: If $W \neq \phi$, then goto step 6.

Step 8: If $i < n$, goto step 1.

Step 9: Apply redundancy elimination on set RS using SURF to generate a set FS.

Step 10: Stop

FS represents the final summary or storyboard of the video. A comparison between the summaries generated using the AMST (AMST) method with the normal MST (MST) is made to evaluate the approach. The total time complexity for the proposed approach using MST may be computed as follows.

Let the total time taken for computing the distance between two frames be t_c. Thus, the total time (T_{MST}) required to generate the MST for one shot $= ({}^{k_i}C_2) \times t_c$, where k_i is the number of frames in the i-th shot S_i of the video V. The total time required for generating the MSTs for all n component shots is

$$T_{\text{MST}} = \sum_{i=1}^{n} ({}^{k_i}C_2) \times t_c \tag{2.10}$$

The total time taken for the AMST approach depends on the number of distances chosen (s), which is a function of the user input percentage (p). The number of distances chosen is given by

$$s = \lceil (p/100) \cdot ({}^kC_2) \rceil \tag{2.11}$$

The time (T_{S_i}) required to generate the AMST for the i-th shot is

$$T_{S_i} = \lceil (p/100) \cdot ({}^{k_i}C_2) \rceil \times t_c + ({}^{k_i}C_2 - s) \times t_p \tag{2.12}$$

where t_p is the prediction time of one distance. The time (T_{AMST}) required to generate the AMST for all n shots is

$$T_{\text{AMST}} = \sum_{i=1}^{n} \lceil (p/100) \cdot ({}^{k_i}C_2) \rceil \times t_c + ({}^{k_i}C_2 - s) \times t_p \tag{2.13}$$

2.4.3 Generation of time-sequenced frames from input video and detection of shot boundaries

The proposed method of video summarization is evaluated over a set of five videos. The input video is first decomposed into its constituent frames by a codec that can decode formats such as, for example, .mp4, .avi and .mpeg. Each video has a frame speed varying between 24 and 30 frames per second. The extracted frames are stored as bitmaps for further analysis. Shot boundaries were detected using the method developed in [8].

2.4.4 User input for percentage of frames to be chosen

For a given set of k frames within a shot, the total number of frame pairs is ${}^{k}C_2$. Computing a large number of correlations results in a huge time complexity. As such, the user is asked for input regarding the percentage of frame pairs to be chosen, which determines the number of distances to be computed. The remaining distances are estimated using interpolation based on these actual distances. The estimation time of the remaining distances is much shorter than the actual distance computation procedure. However, it is to be noted here that a minimum number of frame distances must be chosen in order to ensure that the error in prediction is not high and within a tolerable range. Considering k frames in a shot, the frame pairs f_i and f_j are such that $\min(|i-j|) = 1$ and $\max(|i-j|) = k-1$. The value $\alpha = |i-j|$ is referred to as the frame distance. To get a good estimate, if at least one distance value is taken from each frame distance, a minimum of $k-1$ distances are to be chosen. This results in a lower bound for the value of p, given by the following equation as

$$p = \lceil (k-1)/({}^{k}C_2) \rceil \tag{2.14}$$

This ensures that at least one distance will be chosen from each frame distance. The number of distances selected is given by

$$s = \lceil (p/100) \cdot ({}^{k}C_2) \rceil \tag{2.15}$$

The remaining ${}^{k}C_2 - s$ distances are predicted from these s distances.

2.4.5 Determination of frame pairs to be chosen for computing correlation

After a video is decomposed into time-sequenced frames, shot boundaries are detected using the algorithm devised in [8]. As the AMST is to be generated for all shots individually, a method for choosing the appropriate distances must be devised. For a shot consisting of k frames, the user shall provide as input the percentage of distances to be chosen from ${}^{k}C_2$ possible distances. The number of distances to be chosen (s) is determined using Equation (2.15). The minimum percentage to be chosen for practical purposes is given by Equation (2.14). The number of distances chosen for each frame distance ($\alpha = 1, 2, 3, \ldots, k-1$) is determined by taking one distance for each possible α in round-robin fashion until the sum total of all distances taken is equal to s. However, for the best possible prediction of the remaining distances using interpolation, it is prudent to take the distance values for a particular α at equally spaced intervals. The interval (β_α) between two known distances is given by $\beta_\alpha = (k - \alpha)/t_\alpha$, where t_α is the number of distances to be chosen for frame distance α. Thus, $s = \sum_{\alpha=1}^{k-1} \beta_\alpha$. If the value of β_α is a floating point number, the method divides the number into two parts. One is the integer part, the other the residue part. The value of the residue is computed as follows:

$$\text{residue} = \beta_\alpha - \lfloor \beta_\alpha \rfloor \tag{2.16}$$

To choose the next frame pair f_{ij}, $i = \lceil(\beta_\alpha + \text{residue})\rceil$ and $j = i + \alpha$. The new residue value is set at $\text{residue}_{\text{new}} = \beta_\alpha + \text{residue}_{\text{old}}$. The process is repeated till all required frame pairs have been chosen. The correlation between chosen frame pairs is computed as explained in Section 2.3.2.

2.4.6 Prediction of missing correlation values and error calculation

In this work, the prediction of missing correlation values is computed using both linear and polynomial interpolation methods, as explained in Sections 2.3.5.1 and 2.3.5.2. The prediction process reduces the time complexity to a great extent. The root-mean-square error is computed by comparing the predicted correlation value with the actual one. It is obvious that if the percentage of actual distances is less, the prediction process will be more prone to error than the case when more actual distances are chosen. However, a trade-off is required between the permissible error and the allowable time complexity. In other words, the percentage of actual distances chosen is inversely proportional to the error induced and directly proportional to the computational time of AMST. Figure 2.6 shows the error induced using both interpolation methods as the percentage of actual distances chosen is increased.

2.4.7 Computation of AMST from interpolation value

To construct an AMST (AMST) for each shot of a video, the correlation value obtained is converted to distance between frame pairs. Since the distance is inversely proportional to the correlation, the distance between frames is computed as follows:

$$\text{distance} = \frac{1}{\text{correlation}}$$

All distances (actual and predicted) are stored in a matrix for computation of the AMST using Prim's algorithm [114].

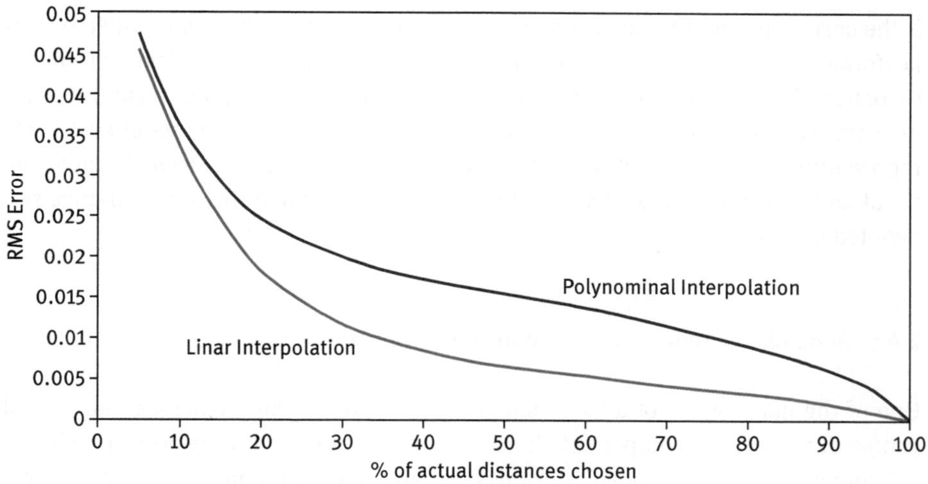

Fig. 2.6: Error induced in AMST due to prediction

2.4.8 Determination of keyframes using AMST

To detect the keyframes within a shot, each frame must be assigned weights so that the frames with the highest weight may be selected as keyframes. Using the three-sigma rule mentioned in Section 2.3.3, a threshold (γ) is to be determined that will serve to discriminate between visually similar and dissimilar frames. After generating the AMST, the average length (μ) and standard deviation of the edges (σ), which constitute the AMST, are computed. The threshold is taken as $\gamma = \mu + 3\sigma$. As shown in Figure 2.7, a disc of radius γ is considered around each frame. The number of frames in the vicinity of a candidate frame is taken as the density of the candidate point. The frames lying on the disc have a visual similarity to the candidate frame. The density of each frame can be taken as its weight. Thus, a matrix consisting of the frame number and its corresponding density is formed. Intuitively, the frame with the highest weight

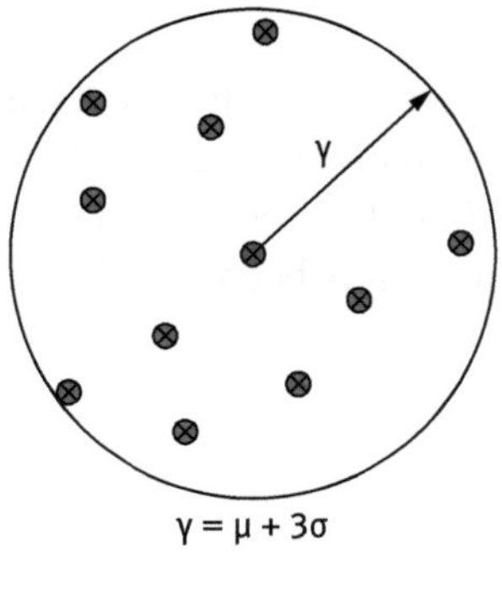

Fig. 2.7: Density computation of a frame

is the best representative and is treated as a keyframe. The frames represented by this keyframe are removed from the matrix. The frame with the highest density from the set of frames left in the matrix is taken as the next keyframe. This process continues till no frames are left to be represented. It can be inferred that the number of keyframes representing a shot is not fixed and depends upon the number of clusters formed. The set of keyframes obtained from the shots in a video is called the representative set, denoted by (RS).

2.4.9 Redundancy reduction on a representative set

One of the major goals of a video summary is to ensure conciseness. In almost all genres of videos, the component shots have redundancies with respect to their visual content. Thus, the removal of frames containing visual redundancies is of prime importance for ensuring the conciseness of the storyboard representation. One of the methods for removing redundant frames is by using mid-level features. Mid-level features capture the semantic content of images to a certain extent by choosing points on an image and representing them by means of feature descriptors. One of the commonly used feature descriptors is SURF [115]. SURF correspondences between a pair of images having similar visual content are shown in Figure 2.8. After generating the set of keyframes over all shots, the redundant keyframes are eliminated using a SURF feature descriptor.

Fig. 2.8: SURF correspondences between two similar images

The number of SURF descriptor matches between a pair of images is an indicator of the similarity in visual content in the images. A high value of similarity indicates redundant frames and results in the elimination of such frames from RS. The redundancy elimination process continues till all pairs of frames within RS have been tested. The final summary is represented by the set FS.

2.4.10 Comparison between static summary generated by AMST method and MST-based approach

To compare the summaries generated using MST and AMST, the system-generated summaries (SGSs) from both methods are compared with the user summaries (USs) generated by ten users. The consensus from all ten USs is known as the final user summary (FUS). The FUS is matched with the SGS to assess the performance of the video summarization method. Two frames are judged to be matching in visual content if the correlation between them is 0.5 or more and the same has been substantiated experimentally. However, in the present work, a higher threshold is used to judge the similarity between two frames. The greater the number of matches between the FUS and SGS, the higher the system effectiveness. The efficiency of the system is determined by taking into consideration two parameters, recall (R) and precision (P), which are defined as follows:

$$\text{Recall} = \frac{(\text{FUS} \cap \text{SGS})}{(\text{FUS})}$$

$$\text{Precision} = \frac{(\text{FUS} \cap \text{SGS})}{(\text{SGS})}$$

Here,
FUS: Set of frames in US
SGS: Set of frames in SGS

2.5 Experimental results and analysis

2.5.1 Video data set

The data set employed to evaluate the proposed method using an AMST consists of five videos. The first video (labelled V1 in Table 2.2) is the Wimbledon Semifinal Highlights of 2013 between Novak Djokovic and Juan Martin Del Potro. The second one is a violin track by "Lindsey Stirling" (labelled V2 in Table 2.2). The third video is "WakaWaka (This Time for Africa)", the official song of the 2010 FIFA World Cup performed by Shakira (marked V3 in Table 2.2). The fourth sequence (labelled V4 in Table 2.2) is another Hindi film song, "Chammak Challo", from the movie "Ra.One". The fifth video

Tab. 2.2: Test video data set

Videos	V1	V2	V3	V4	V5
Duration (mm:ss)	02:58	03:27	03:31	04:10	05:45
No. of frames	4468	4965	5053	6265	8632
No. of hard cuts	43	76	137	159	62
Average no. of frames in each shot	101.54	64.48	36.61	39.16	137.02

(labelled V5 in Table 2.2) is also a movie song, "You Are My Soniya", from the Hindi film "Khabhi Khusi Khabi Gham". The videos chosen for experimentation have varied characteristics.

All the videos used for this work were downloaded from YouTube. The resolution of all the videos in the data set is 640 × 360 pixels at a frame rate of 25 fps. All videos in the data set are in MP4 file format, commonly called MPEG-4 file format version 2. Table 2.2 lists the various characteristics of the videos in terms of time duration (mm:ss), number of frames in each video, number of hard cuts and average number of frames in each shot.

2.5.2 Experimental results

The major aim of the AMST-based method was to reduce the time complexity involved in the MST approach. As such, to obtain a fair idea of the time taken in both approaches, a comparison of the actual time taken to generate the MST and that required for the AMST was made. The cumulative time (in seconds) is plotted with the progression of each shot as shown in Figure 2.9. Here, it was assumed that the the percentage

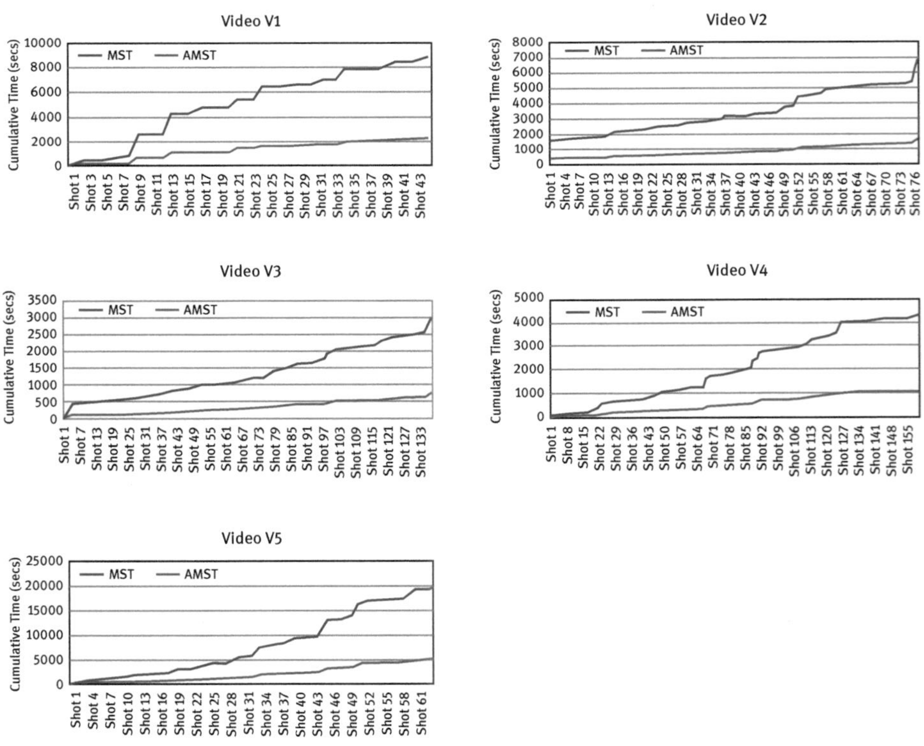

Fig. 2.9: Time comparison between MST and AMST

of actual distances chosen would be $p = 25\%$ and 75% of the distances would be predicted from the given actual distances using interpolation. Figure 2.9 demonstrates the effectiveness of the proposed method at saving huge amounts of time to generate the storyboard, largely without sacrificing the quality of summarization.

To evaluate the AMST method, the recall and precision values are computed for different percentages of the actual distances. It is obvious that both these metrics reach their highest values at 100% of the actual distances, in which case the AMST and MST are identical. A graphical plot of the recall and precision values for linear and polynomial interpolation are shown in Figure 2.10 and 2.11 respectively for different percentages of the actual distances chosen. As expected, both metrics increase for increased values of the number of actual distances chosen and as the number of predicted values is progressively reduced. Since the number of predicted values decreases with increases in the percentage of actual distances, the AMST generated is prone to less error. In some cases, such as between 5% and 10% in linear interpolation for Video 3, 25% to 30% in polynomial interpolation for Video 2, and so forth, it is observed that the values of recall and precision decrease for a small interval even when the number of actual distances chosen is increased.

2.6 Conclusion

The work presented in this chapter was aimed at reducing the time complexity of the MST-based method for static video summarization. It is seen that by computing a smaller number of distances and predicting the remaining ones using interpolation, the performance of the system in terms of recall and precision is not degraded to a large extent. However, as a future expansion of the present work, a mathematical model may be developed that will be able to provide the amount of error (both minimum and maximum) that would be induced during the prediction phase. Since the proposed approach tries to find the optimum set of keyframes through the generation of an approximate MST, it is still not appropriate for real-time applications where the selection of keyframes must be made on the fly and in a time frame that is much smaller than the duration of the video.

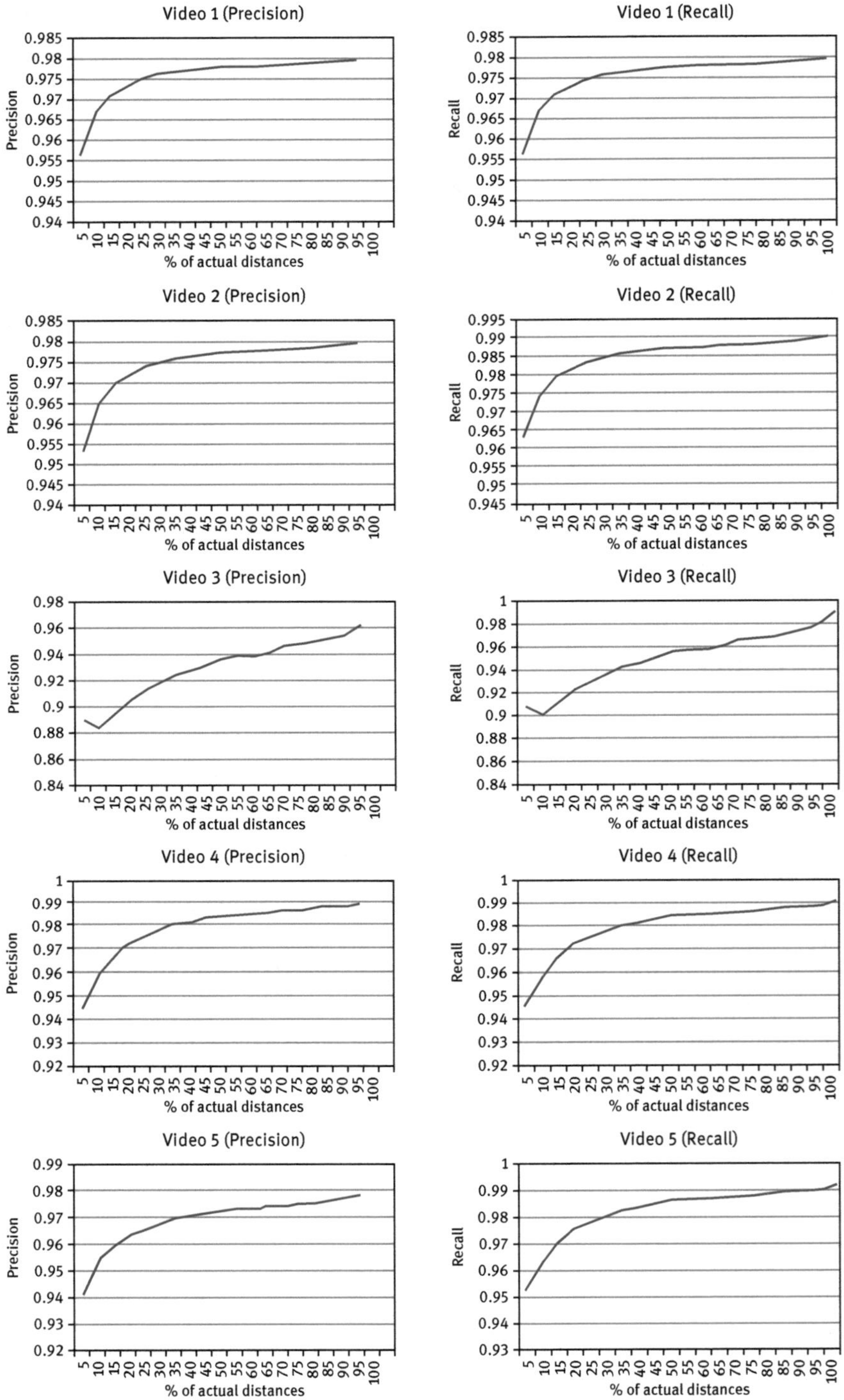

Fig. 2.10: Recall and precision for different percentages of actual distances based on linear interpolation

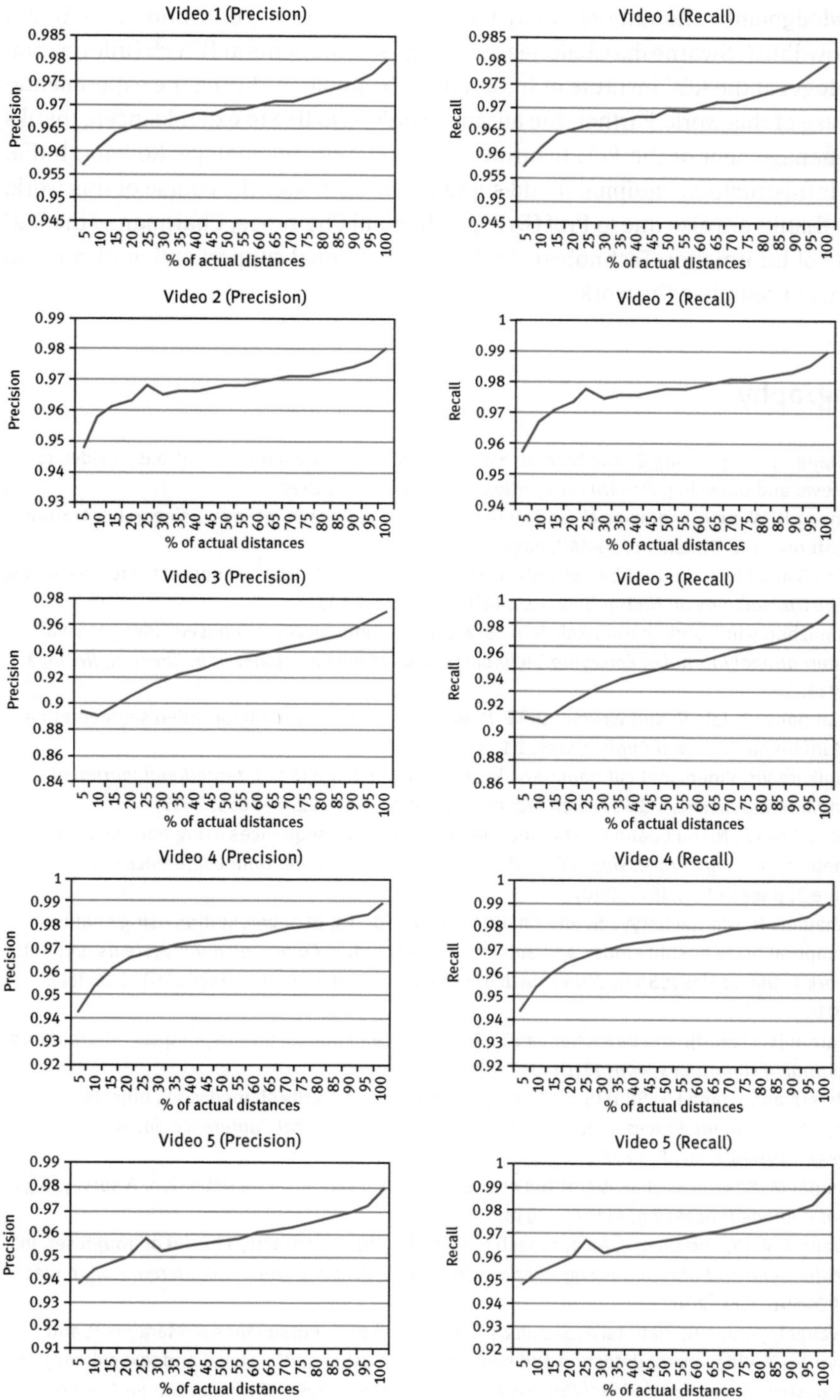

Fig. 2.11: Recall and precision for different percentages of actual distances based on polynomial interpolation

Acknowledgment: The authors would like to acknowledge the contributions of Ankita Roy Chowdhury, Swarnesha Chaterjee and Gargi Sau, students at B.Tech (Information Technology) at the RCC Institute of Information Technology, during the experimentation phase of this work. Further, the authors would also like to extend sincere thanks to the management of the RCC Institute of Information Technology, Kolkata, India, for the infrastructure, equipment and support provided over the course of this work. Sincere thanks are also due to Prof (Dr) Sparshamani Chatterjee, Chairman of the RCC Institute of Information Technology, for his unconditional support and motivation to enable us to complete the work.

Bibliography

[1] Zhang HJ, Wu J, Zhong D, and Smoliar SW. An integrated system for content-based video retrieval and browsing. *Pattern recognition*, 30(4):643–658, 1997.

[2] Zhang D, Wong A, Indrawan M, and Lu G. Content-based image retrieval using gabor texture features. *IEEE Transactions PAMI*, pages 13–15, 2000.

[3] Guo G and Li SZ. Content-based audio classification and retrieval by support vector machines. *IEEE transactions on Neural Networks*, 14(1):209–215, 2003.

[4] Tamizharasan C and Chandrakala S. A survey on multimodal content based video retrieval. *International Journal of Emerging Technology and Advanced Engineering. Chennai, INDIA*, 3, 2013.

[5] Hampapur A, Jain R, and Weymouth TE. Production model based digital video segmentation. *Multimedia Tools and Applications*, 1(1):9–46, 1995.

[6] Thakore VH. Video shot cut boundary detection using histogram. *International Journal of Engineering Sciences & Research Technology, ISSN*, pages 2277–9655, 2013.

[7] Cernekova Z, Nikou C, and Pitas I. Shot detection in video sequences using entropy based metrics. In *Image Processing. 2002. Proceedings. 2002 International Conference on*, volume 3, pages III–III. IEEE, 2002.

[8] Bhaumik H, Bhattacharyya S, and Chakraborty S. Video shot segmentation using spatio-temporal fuzzy hostility index and automatic threshold. In *Communication Systems and Network Technologies (CSNT), 2014 Fourth International Conference on*, pages 501–506. IEEE, 2014.

[9] Barron JL, Fleet DJ, and Beauchemin SS. Performance of optical flow techniques. *International journal of computer vision*, 12(1):43–77, 1994.

[10] Ostermann J, Jang ES, Shin JS, and Chen T. Coding of arbitrarily shaped video objects in mpeg-4. In *Image Processing, 1997. Proceedings., International Conference on*, volume 1, pages 496–499. IEEE, 1997.

[11] Kumar M, Saxena R, et al. Algorithm and technique on various edge detection: A survey. *Signal & Image Processing*, 4(3):65, 2013.

[12] Wang T, Wu Y, and Chen L. An approach to video key-frame extraction based on rough set. In *Multimedia and Ubiquitous Engineering, 2007. MUE'07. International Conference on*, pages 590–596. IEEE, 2007.

[13] Evangelopoulos G, Zlatintsi A, Skoumas G, Rapantzikos K, Potamianos A, Maragos P, and Avrithis Y. Video event detection and summarization using audio, visual and text saliency. In *Acoustics, Speech and Signal Processing, 2009. ICASSP 2009. IEEE International Conference on*, pages 3553–3556. IEEE, 2009.

[14] Orrite C, Martinez F, Herrero E, Ragheb H, and Velastin S. Independent viewpoint silhouette-based human action modelling and recognition. In *The 1st International Workshop on Machine Learning for Vision-based Motion Analysis-MLVMA'08*, 2008.
[15] Joshi KA and Thakore DG. A survey on moving object detection and tracking in video surveillance system. *International Journal of Soft Computing and Engineering*, 2(3):44–48, 2012.
[16] Poppe R. A survey on vision-based human action recognition. *Image and vision computing*, 28(6):976–990, 2010.
[17] Cohen I, Sebe N, Garg A, Chen LS, and Huang TS. Facial expression recognition from video sequences: temporal and static modeling. *Computer Vision and image understanding*, 91(1–2):160–187, 2003.
[18] Bhaumik H, Bhattacharyya S, Das M, and Chakraborty S. Enhancement of perceptual quality in static video summarization using minimal spanning tree approach. In *Signal Processing, Informatics, Communication and Energy Systems (SPICES), 2015 IEEE International Conference on*, pages 1–7. IEEE, 2015.
[19] Bhaumik H, Bhattacharyya S, and Chakraborty S. Redundancy elimination in video summarization. In *Image Feature Detectors and Descriptors*, pages 173–202. Springer, 2016.
[20] Mundur P, Rao Y, and Yesha Y. Keyframe-based video summarization using delaunay clustering. *International Journal on Digital Libraries*, 6(2):219–232, 2006.
[21] De Avila SEF, Lopes APB, da Luz A, and de Albuquerque Araújo A. Vsumm: A mechanism designed to produce static video summaries and a novel evaluation method. *Pattern Recognition Letters*, 32(1):56–68, 2011.
[22] Hadi Y, Essannouni F, and Thami RH. Video summarization by k-medoid clustering. In *Proceedings of the 2006 ACM symposium on Applied computing*, pages 1400–1401. ACM, 2006.
[23] Hanjalic A and Zhang H. An integrated scheme for automated video abstraction based on unsupervised cluster-validity analysis. *IEEE Transactions on circuits and systems for video technology*, 9(8):1280–1289, 1999.
[24] Cahuina EJYC and Chavez GC. A new method for static video summarization using local descriptors and video temporal segmentation. In *Graphics, Patterns and Images (SIBGRAPI), 2013 26th SIBGRAPI-Conference on*, pages 226–233. IEEE, 2013.
[25] Bhaumik H, Bhattacharyya S, Dutta S, and Chakraborty S. Towards redundancy reduction in storyboard representation for static video summarization. In *Advances in Computing, Communications and Informatics (ICACCI), 2014 International Conference on*, pages 344–350. IEEE, 2014.
[26] Chang HS, Sull S, and Lee SU. Efficient video indexing scheme for content-based retrieval. *IEEE Transactions on Circuits and Systems for Video Technology*, 9(8):1269–1279, 1999.
[27] DeMenthon D, Kobla V, and Doermann D. Video summarization by curve simplification. In *Proceedings of the sixth ACM international conference on Multimedia*, pages 211–218. ACM, 1998.
[28] Yeung MM and Yeo BL. Video visualization for compact presentation and fast browsing of pictorial content. *IEEE Transactions on circuits and Systems for Video Technology*, 7(5):771–785, 1997.
[29] Fränti P, Mednonogov A, Kyrki V, and Kälviäinen H. Content-based matching of line-drawing images using the hough transform. *International Journal on Document Analysis and Recognition*, 3(2):117–124, 2000.
[30] Gygli M, Grabner H, and Van Gool L. Video summarization by learning submodular mixtures of objectives. In *Proceedings CVPR 2015*, pages 3090–3098, 2015.
[31] Xu J, Mukherjee L, Li Y, Warner J, Rehg JM, and Singh V. Gaze-enabled egocentric video summarization via constrained submodular maximization. In *Computer Vision and Pattern Recognition (CVPR), 2015 IEEE Conference on*, pages 2235–2244. IEEE, 2015.

[32] Itazuri T, Fukusato T, Yamaguchi S, and Morishima S. Court-based volleyball video summarization focusing on rally scene. In *Proceedings of the IEEE Conference on Computer Vision and Pattern Recognition Workshops*, pages 123–130, 2017.

[33] Zlatintsi A, Koutras P, Evangelopoulos G, Malandrakis N, Efthymiou N, Pastra K, Potamianos A, and Maragos P. Cognimuse: a multimodal video database annotated with saliency, events, semantics and emotion with application to summarization. *EURASIP Journal on Image and Video Processing*, 2017(1):54, 2017.

[34] Gharbi H, Bahroun S, Massaoudi M, and Zagrouba E. Key frames extraction using graph modularity clustering for efficient video summarization. In *Acoustics, Speech and Signal Processing (ICASSP), 2017 IEEE International Conference on*, pages 1502–1506. IEEE, 2017.

[35] Sharghi A, Laurel JS, and Gong B. Query-focused video summarization: Dataset, evaluation, and a. memory network based approach. In *2017 IEEE Conference on Computer Vision and Pattern Recognition (CVPR)*, pages 2127–2136. IEEE, 2017.

[36] Lin ST, Liao YH, Tsao Y, and Chien SY. Object-based on-line video summarization for internet of video things. In *2017 IEEE International Symposium on Circuits and Systems (ISCAS)*, pages 1–4. IEEE, 2017.

[37] Chen X, Zhang Y, Ai Q, Xu H, Yan J, and Qin Z. Personalized key frame recommendation. In *Proceedings of the 40th International ACM SIGIR Conference on Research and Development in Information Retrieval*, pages 315–324. ACM, 2017.

[38] Sahoo A, Kaushal V, Doctor K, Shetty S, Iyer R, and Ramakrishnan G. A unified multi-faceted video summarization system. *arXiv preprint arXiv:1704.01466*, 2017.

[39] Mirzasoleiman B, Jegelka S, and Krause A. Streaming non-monotone submodular maximization: Personalized video summarization on the fly. *arXiv preprint arXiv:1706.03583*, 2017.

[40] Choi J, Oh TH, and Kweon IS. Textually customized video summaries. *arXiv preprint arXiv:1702.01528*, 2017.

[41] Otani M, Nakashima Y, Sato T, and Yokoya N. Video summarization using textual descriptions for authoring video blogs. *Multimedia Tools and Applications*, 76(9):12097–12115, 2017.

[42] Mitra A, Biswas S, and Bhattacharyya C. Bayesian modeling of temporal coherence in videos for entity discovery and summarization. *IEEE transactions on pattern analysis and machine intelligence*, 39(3):430–443, 2017.

[43] Mademlis I, Tefas A, and Pitas I. Summarization of human activity videos using a salient dictionary. In *Proceedings of the IEEE International Conference on Image Processing (ICIP)*, 2017.

[44] Panda R and Roy-Chowdhury AK. Collaborative summarization of topic-related videos. In *CVPR*, volume 2, no. 4, page 5, 2017.

[45] Abdelali AB, Hannachi M, Touil L, and Mtibaa A. Adequation and hardware implementation of the color structure descriptor for real-time temporal video segmentation. *Journal of Real-Time Image Processing*, 13(4):739–758, 2017.

[46] del Molino AG, Tan C, Lim JH, and Tan AH. Summarization of egocentric videos: A comprehensive survey. *IEEE Transactions on Human-Machine Systems*, 47(1):65–76, 2017.

[47] Ji Z, Xiong K, Pang Y, and Li X. Video summarization with attention-based encoder-decoder networks. *arXiv preprint arXiv:1708.09545*, 2017.

[48] del Molino AG, Mandal B, Lin J, Lim JH, Subbaraju V, and Chandrasekhar V. VC-I2R@ ImageCLEF2017: ensemble of deep learned features for lifelog video summarization. In *CLEF working notes, CEUR*, pages 11–14, 2017.

[49] Panda R, Mithun NC, and Roy-Chowdhury AK. Diversity-aware multi-video summarization. *IEEE Transactions on Image Processing*, 26(10):4712–4724, 2017.

[50] Panda R and Roy-Chowdhury AK. Multi-view surveillance video summarization via joint embedding and sparse optimization. *IEEE Transactions on Multimedia*, 19(9):2010–2021, 2017.

[51] Meng J, Wang S, Wang H, Yuan J, and Tan YP. Video summarization via multi-view representative selection. *IEEE transactions on image processing: a publication of the IEEE Signal Processing Society*, 27(5):2134–2145, 2018.

[52] Zhang Y, Liang X, Zhang D, Tan M, and Xing EP. Unsupervised object-level video summarization with online motion auto-encoder. *arXiv preprint arXiv:1801.00543*, 2018.

[53] Rudinac S, Chua TS, Diaz-Ferreyra N, Friedland G, Gornostaja T, Huet B, Kaptein R, Lindén K, Moens MF, Peltonen J, et al. Rethinking summarization and storytelling for modern social multimedia. *Lecture Notes in Computer Science*, 10704, 2018.

[54] Tejero-de Pablos A, Nakashima Y, Sato T, Yokoya N, Linna M, and Rahtu E. Summarization of user-generated sports video by using deep action recognition features. *IEEE Transactions on Multimedia*, 2018.

[55] Plummer BA, Brown M, and Lazebnik S. Enhancing video summarization via vision-language embedding. In *Computer Vision and Pattern Recognition*, 2017.

[56] Fang H, Jiang J, and Feng Y. A fuzzy logic approach for detection of video shot boundaries. *Pattern Recognition*, 39(11):2092–2100, 2006.

[57] Das S, Sural S, and Majumdar AK. Detection of hard cuts and gradual transitions from video using fuzzy logic. *International Journal of Artificial Intelligence and Soft Computing*, 1(1):77–98, 2008.

[58] Anderson D, Luke RH, Keller JM, Skubic M, Rantz M, and Aud M. Linguistic summarization of video for fall detection using voxel person and fuzzy logic. *Computer vision and image understanding*, 113(1):80–89, 2009.

[59] Cayllahua-Cahuina E, Cámara-Chávez G, and Menotti D. A static video summarization approach with automatic shot detection using color histograms. In *Proceedings of the International Conference on Image Processing, Computer Vision, and Pattern Recognition (IPCV)*, page 1. The Steering Committee of The World Congress in Computer Science, Computer Engineering and Applied Computing (WorldComp), 2012.

[60] Carpenter GA, Grossberg S, and Rosen DB. Fuzzy art: Fast stable learning and categorization of analog patterns by an adaptive resonance system. *Neural networks*, 4(6):759–771, 1991.

[61] Bezdek JC, Ehrlich R, and Full W. Fcm: The fuzzy c-means clustering algorithm. *Computers & Geosciences*, 10(2–3):191–203, 1984.

[62] Doulamis AD, Doulamis ND, and Kollias SD. A fuzzy video content representation for video summarization and content-based retrieval. *Signal Processing*, 80(6):1049–1067, 2000.

[63] Peng WT, Chu WT, Chang CH, Chou CN, Huang WJ, Chang WY, and Hung YP. Editing by viewing: automatic home video summarization by viewing behavior analysis. *IEEE Transactions on Multimedia*, 13(3):539–550, 2011.

[64] Elbouz M, Alfalou A, and Brosseau C. Fuzzy logic and optical correlation-based face recognition method for patient monitoring application in home video surveillance. *Optical Engineering*, 50(6):067003–067003, 2011.

[65] Zhou Z, Chen X, Chung YC, He Z, Han TX, and Keller JM. Activity analysis, summarization, and visualization for indoor human activity monitoring. *IEEE Transactions on Circuits and Systems for Video Technology*, 18(11):1489–1498, 2008.

[66] Avrithis YS, Doulamis AD, Doulamis ND, and Kollias SD. A stochastic framework for optimal key frame extraction from mpeg video databases. *Computer Vision and Image Understanding*, 75(1–2):3–24, 1999.

[67] Doulamis N, Doulamis A, and Ntalianis K. An optimal interpolation-based scheme for video summarization. In *Multimedia and Expo, 2002. ICME'02. Proceedings. 2002 IEEE International Conference on*, volume 1, pages 297–300. IEEE, 2002.

[68] Karpathy A, Toderici G, Shetty S, Leung T, Sukthankar R, and Fei-Fei L. Large-scale video classification with convolutional neural networks. In *Proceedings of the IEEE conference on Computer Vision and Pattern Recognition*, pages 1725–1732, 2014.

[69] Yao L, Torabi A, Cho K, Ballas N, Pal C, Larochelle H, and Courville A. Describing videos by exploiting temporal structure. In *Proceedings of the IEEE international conference on computer vision*, pages 4507–4515, 2015.

[70] Venugopalan S, Xu H, Donahue J, Rohrbach M, Mooney R, and Saenko K. Translating videos to natural language using deep recurrent neural networks. *arXiv preprint arXiv:1412.4729*, 2014.

[71] Doulamis A, Doulamis N, Ntalianis K, and Kollias S. An efficient fully unsupervised video object segmentation scheme using an adaptive neural-network classifier architecture. *IEEE Transactions on Neural Networks*, 14(3):616–630, 2003.

[72] Zawbaa HM, El-Bendary N, Hassanien AE, and Kim Th. Machine learning-based soccer video summarization system. *Multimedia, Computer Graphics and Broadcasting*, pages 19–28, 2011.

[73] Panda R, Das A, Wu Z, Ernst J, and Roy-Chowdhury AK. Weakly supervised summarization of web videos. In *2017 IEEE International Conference on Computer Vision (ICCV)*, pages 3677–3686. IEEE, 2017.

[74] Sah S, Kulhare S, Gray A, Venugopalan S, Prud'Hommeaux E, and Ptucha R. Semantic text summarization of long videos. In *Applications of Computer Vision (WACV), 2017 IEEE Winter Conference on*, pages 989–997. IEEE, 2017.

[75] Sanghavi RS and Pangaonkar SA. An approach to detect indoor outdoor scene for effective video summarization. *International Journal of Engineering Technology, Management and Applied Sciences*, 5(6):976–981, 2017.

[76] Chiu P, Girgensohn A, Polak W, Rieffel E, and Wilcox L. A genetic algorithm for video segmentation and summarization. In *Multimedia and Expo, 2000. ICME 2000. 2000 IEEE International Conference on*, volume 3, pages 1329–1332. IEEE, 2000.

[77] Torres RdS, Falcão AX, Gonçalves MA, Papa JP, Zhang B, Fan W, and Fox EA. A genetic programming framework for content-based image retrieval. *Pattern Recognition*, 42(2):283–292, 2009.

[78] Lu J, Zhao T, and Zhang Y. Feature selection based-on genetic algorithm for image annotation. *Knowledge-Based Systems*, 21(8):887–891, 2008.

[79] Yoo HW and Cho SB. Video scene retrieval with interactive genetic algorithm. *Multimedia Tools and Applications*, 34(3):317–336, 2007.

[80] Mademlis I, Tefas A, Nikolaidis N, and Pitas I. Summarization of human activity videos via low-rank approximation. In *Acoustics, Speech and Signal Processing (ICASSP), 2017 IEEE International Conference on*, pages 1627–1631. IEEE, 2017.

[81] Ng HW, Sawahata Y, and Aizawa K. Summarization of wearable videos using support vector machine. In *Multimedia and Expo, 2002. ICME'02.Proceedings. 2002 IEEE International Conference on*, volume 1, pages 325–328. IEEE, 2002.

[82] Li L, Zhou K, Xue GR, Zha H, and Yu Y. Video summarization via transferrable structured learning. In *Proceedings of the 20th international conference on World wide web*, pages 287–296. ACM, 2011.

[83] Potapov D, Douze M, Harchaoui Z, and Schmid C. Category-specific video summarization. In *ECCV 2014-European Conference on Computer Vision*, volume 8694, pages 540–555. Springer, 2014.

[84] Ling X, Yuanxin O, Huan L, and Zhang X. A method for fast shot boundary detection based on svm. In *Image and Signal Processing, 2008. CISP'08. Congress on*, volume 2, pages 445–449. IEEE, 2008.

[85] Sadlier DA and O'Connor NE. Event detection in field sports video using audio-visual features and a support vector machine. *IEEE Transactions on Circuits and Systems for Video Technology*, 15(10):1225–1233, 2005.
[86] Wu K and Yap KH. Fuzzy svm for content-based image retrieval: a pseudo-label support vector machine framework. *IEEE Computational Intelligence Magazine*, 1(2):10–16, 2006.
[87] Ngo CW, Ma YF, and Zhang HJ. Video summarization and scene detection by graph modeling. *IEEE Transactions on Circuits and Systems for Video Technology*, 15(2):296–305, 2005.
[88] Ngo CW, Ma YF, and Zhang HJ. Automatic video summarization by graph modeling. In *Computer Vision, 2003. Proceedings. Ninth IEEE International Conference on*, pages 104–109. IEEE, 2003.
[89] Lu S, Lyu MR, and King I. Video summarization by spatial-temporal graph optimization. In *Circuits and Systems, 2004. ISCAS'04. Proceedings of the 2004 International Symposium on*, volume 2, pages II–197. IEEE, 2004.
[90] Peyrard N and Bouthemy P. Motion-based selection of relevant video segments for video summarization. *Multimedia Tools and Applications*, 26(3):259–276, 2005.
[91] Piriou G, Bouthemy P, and Yao JF. Recognition of dynamic video contents with global probabilistic models of visual motion. *IEEE Transactions on Image Processing*, 15(11):3417–3430, 2006.
[92] Ma YF, Lu L, Zhang HJ, and Li M. A user attention model for video summarization. In *Proceedings of the tenth ACM international conference on Multimedia*, pages 533–542. ACM, 2002.
[93] Gong Y and Liu X. Video summarization using singular value decomposition. In *Computer Vision and Pattern Recognition, 2000. Proceedings. IEEE Conference on*, volume 2, pages 174–180. IEEE, 2000.
[94] Orriols X and Binefa X. An em algorithm for video summarization, generative model approach. In *Computer Vision, 2001. ICCV 2001.Proceedings. Eighth IEEE International Conference on*, volume 2, pages 335–342. IEEE, 2001.
[95] Wang M, Hong R, Li G, Zha ZJ, Yan S, and Chua TS. Event driven web video summarization by tag localization and key-shot identification. *IEEE Transactions on Multimedia*, 14(4):975–985, 2012.
[96] Boreczky JS and Rowe LA. Comparison of video shot boundary detection techniques. *Journal of Electronic Imaging*, 5(2):122–128, 1996.
[97] Hu W, Xie N, Li L, Zeng X, and Maybank S. A survey on visual content-based video indexing and retrieval. *IEEE Transactions on Systems, Man, and Cybernetics, Part C (Applications and Reviews)*, 41(6):797–819, 2011.
[98] Cotsaces C, Nikolaidis N, and Pitas I. Video shot detection and condensed representation. a review. *IEEE signal processing magazine*, 23(2):28–37, 2006.
[99] Bhaumik H, Bhattacharyya S, Nath MD, and Chakraborty S. Hybrid soft computing approaches to content based video retrieval: A brief review. *Applied Soft Computing*, 46:1008–1029, 2016.
[100] Xiong W and Lee JCM. Efficient scene change detection and camera motion annotation for video classification. *Computer vision and image understanding*, 71(2):166–181, 1998.
[101] Koprinska I and Carrato S. Temporal video segmentation: A survey. *Signal processing: Image communication*, 16(5):477–500, 2001.
[102] Snoek CGM and Worring M. Multimodal video indexing: A review of the state-of-the-art. *Multimedia tools and applications*, 25(1):5–35, 2005.
[103] Bhaumik H, Bhattacharyya S, Chakraborty M, and Chakraborty S. Dissolve detection in videos using an ensemble approach. In *Advances in Computing, Communications and Informatics (ICACCI), 2015 International Conference on*, pages 1461–1467. IEEE, 2015.

[104] Song B and Ra JB. Automatic shot change detection algorithm using multi-stage clustering for mpeg-compressed videos. *Journal of Visual Communication and Image Representation*, 12(3):364–385, 2001.

[105] Lee HC, Lee CW, and Kim S. Abrupt shot change detection using an unsupervised clustering of multiple features. In *Acoustics, Speech, and Signal Processing, 2000. ICASSP'00.Proceedings. 2000 IEEE International Conference on*, volume 4, pages 2015–2018. IEEE, 2000.

[106] Bhaumik H, Bhattacharyya S, and Chakraborty S. An unsupervised method for real time video shot segmentation. In *Proceedings of International Conference on Digital Image Processing and Pattern Recognition*, pages 307–318, 2014.

[107] Hanjalic A. Shot-boundary detection: Unraveled and resolved? *IEEE transactions on circuits and systems for video technology*, 12(2):90–105, 2002.

[108] Yuan J, Wang H, Xiao L, Zheng W, Li J, Lin F, and Zhang B. A formal study of shot boundary detection. *IEEE transactions on circuits and systems for video technology*, 17(2):168–186, 2007.

[109] Lienhart R. Comparison of automatic shot boundary detection algorithms. In *Storage and Retrieval for Image and Video Databases (SPIE)*, pages 290–301, 1999.

[110] Petersohn C. Fraunhofer hhi at trecvid 2004: Shot boundary detection system. In *TREC Video Retrieval Evaluation Online Proceedings, TRECVID*, 2004.

[111] Cernekova Z, Pitas I, and Nikou C. Information theory-based shot cut/fade detection and video summarization. *IEEE Transactions on circuits and systems for video technology*, 16(1):82–91, 2006.

[112] Jeong S. Histogram-based color image retrieval. *Psych221/EE362 project report*, 2001.

[113] Patel BV and Meshram BB. Content based video retrieval systems. *arXiv preprint arXiv:1205.1641*, 2012.

[114] Moret BME and Shapiro HD. An empirical analysis of algorithms for constructing a minimum spanning tree. In *Workshop on Algorithms and Data Structures*, pages 400–411. Springer, 1991.

[115] Bay H, Tuytelaars T, and Van Gool L. Surf: Speeded-up robust features. In *Computer vision–ECCV 2006*, pages 404–417. Springer, 2006.

Amit Adate and B. K. Tripathy*

3 Deep learning techniques for image processing

Abstract: Most techniques in image processing involve algorithms that are custom built and lack flexibility, making them different to the data being processed. In this chapter we elaborate upon various methodologies within the domain of image processing. We chronologically demonstrate the role of learning techniques involved in image super resolution, image upsampling, image quality assessment and parallel computing techniques. Further, an in-depth explanation is provided of the involvement of deep neural architectures as an impressive tool for solving multiple image processing problems. This chapter describes the superior performance obtained by the application of deep learning techniques to the task of image processing.

Keywords: Deep learning, Image processing, Neural architectures, Classification, Image resolution, Image sharpening

3.1 Introduction

The world has become an automated one as we move closer towards white collar automation faster than ever before. One critical insight that only people can provide are visual abilities that help them to manoeuvre in real life and provide value to society. Deep learning is playing catch-up in this part of image and vision processing with newer algorithms and better performance aspects. This chapter provides a detailed review of the topic using a sequential time line of advances in the image and video processing power of today's algorithms with a final note on the growing use of parallel computing hardware, which is being used for training these deep neural networks.

The chapter introduces the concept of neural networks and how such networks can be used for training on a variety of different types of data to represent all kinds of policies. It then covers image and video processing techniques like enhancement, upsampling, super resolution and quality assessment of images and how they have advanced in their respective techniques along with how parallel computing and deep learning fit together.

Deep learning is a relatively new field what began with the advent of greater computing power, but the term started to be used in the research community around 1986 when Dechter et al. [1] published a paper introducing a pseudo-deep learning concept to solve constraint specification problems, a niche application for solving constraint specification problems. The approach looks at backtracking search that was used in Prolog for solving dead-end situations and proving truth maintenance system theorem. The researchers showed that it was possible to use backtracking and back-

Amit Adate, VIT University
***Corresponding author: B. K. Tripathy,** VIT University
https://doi.org/10.1515/9783110551433-003

propagation for learning methodology in neural networks [2]. The first official work on learning algorithms was published by Aizenberg et al. in their book [3].

Deep learning uses a layered structure of neurons that take a large data set, propagate the data forwards and backpropagate the results using a loss function and gradient over the connections between the layers, which simulates the learning process [4]. Deep learning has been extremely successful in image classification work with the advent of convolutional neural networks (CNNs) [5]. This architecture involves using a convolutional layer for feature extraction along with maxpooling and fully connected layers for classification.

3.2 Image and video enhancement

Image enhancement generally includes techniques that are used to modify an image or a video in such a way that its quality and visibility increase. This can be done in a number of traditional ways that can be broadly split into two categories: spatial domain and frequency domain.

Spatial domain methods alter an image by directly altering the pixel values in the image using transformations and filters. This direct change allows for less complex yet less sophisticated changes to be made to the image. Frequency domain methods first convert an image from one domain to another domain and then transform the image using various techniques before converting the image to the original domain [6].

Deep learning methods for image enhancement usually apply generative models to use previously learned features and apply them to a noisy image to generate a more aesthetically pleasing image that has enhanced features [7]. Another method involves using a bilateral architecture to take the low-resolution and high-resolution versions of the same image and use a CNN on the low-resolution image to generate coefficients that are later added to the high-resolution version [8].

3.2.1 Pixel processing operations

To conduct operations in the spatial domain, the earliest applications begins in the neighbourhood of the pixels themselves. Because each pixel in an image is surrounded by other pixels, they are subjected to a series of transformations. These transformations themselves enhance the image, but their most popular use is to act as a preprocessing method to assist in making further, complex changes. Some transformations include thresholding transformations, power-law transformations and logarithmic transformations. Specifically for image enhancement, various classes of neural filters have been proposed. A neural filter primarily requires processing an image through two separate classes of a decision-based filter. The output is then passed on to a feedforward neural network.

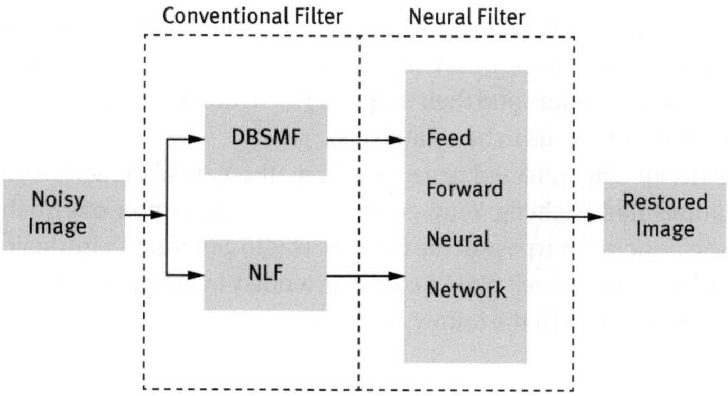

Fig. 3.1: Pixel processing neural unit

The Figure 3.1 displays two filters, a conventional filter and a neural filter. Both together act as an impulse noise removal filter. Initially the noisy image is given to conventional filters like a non-linear filter(NF) and descision based switching median filter (DBSMF). It is then processed further using a simple feedforward neural network (Figure 3.2). The fine edges are preserved and noise is eliminated using a backpropagation algorithm.

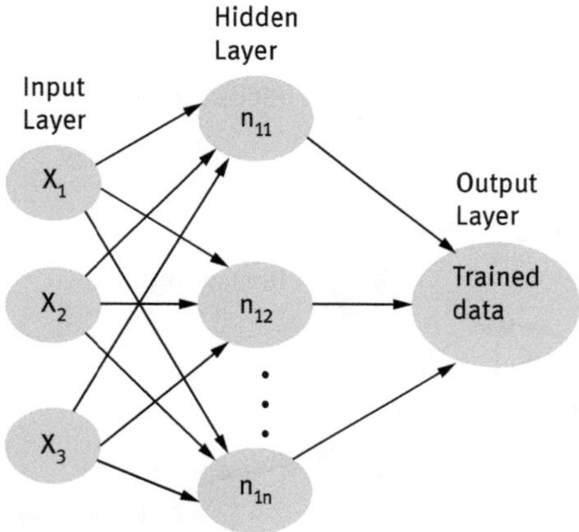

Fig. 3.2: A feedforward neural network

3.2.2 Histogram-based image enhancement

These are contrast enhancement techniques, and every technique is dependent upon a histogram of the image. To perform any operation in the frequency domain, there has to be a representation of the tonal distribution of the image. Hence, after pixel level transformations are undertaken, a computational model is implemented to automatically enhance the image following the learned style.

Learning image enhancements based on its visual style is a challenging task; it is a highly observational and visceral process that deals with the information embedded in an image based upon its pixel colours and their frequency. To learn an enhancement style, a quantitative relationship needs to be established.

Deep neural networks are implemented to generate a mathematical model based upon quantitative relationships. Zicheng Yang et al. [9] elaborated upon one such deep neural network; the principal purpose of the network was to generate an intrinsic colour mapping function that maps each pixel's colour in a noisy to enhanced image. The architecture is elaborated upon in the following figure.

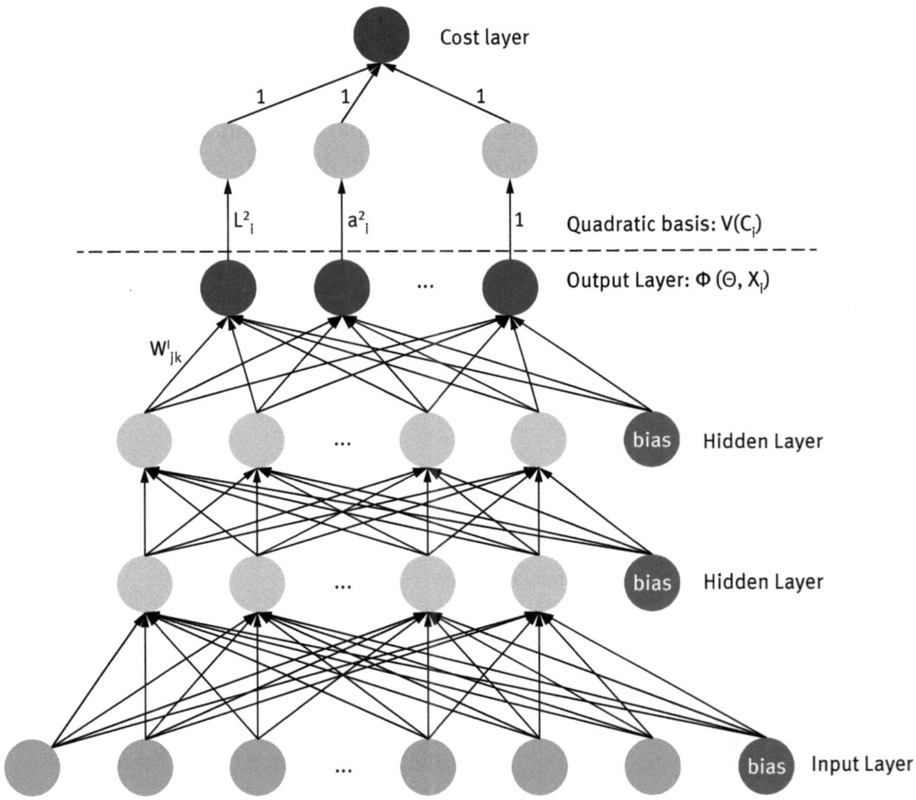

Fig. 3.3: LAPGAN training procedure

In Figure 3.3, the neurons above the dotted line define how the intrinsic function is computed. The weights between the yellow and blue neurons are all components of the quadratic colour basis, and the identity function is the activation function within the purple and yellow neurons.

This architecture is superior to its predecessors as it is able to learn complex and highly non-linear photo enhancement styles. It is capable of enhancing preferred contextual features using the robust regression capability of deep neural networks.

3.3 Image super resolution

Another scenario of image enhancement can also be in image super resolution, where low-resolution images are taken and bumped up to create a higher-resolution version of the same image. This use case has been distinguished from image enhancement as research in this category has been extensive regarding generative models and is beyond the scope of this chapter (Figure 3.4).

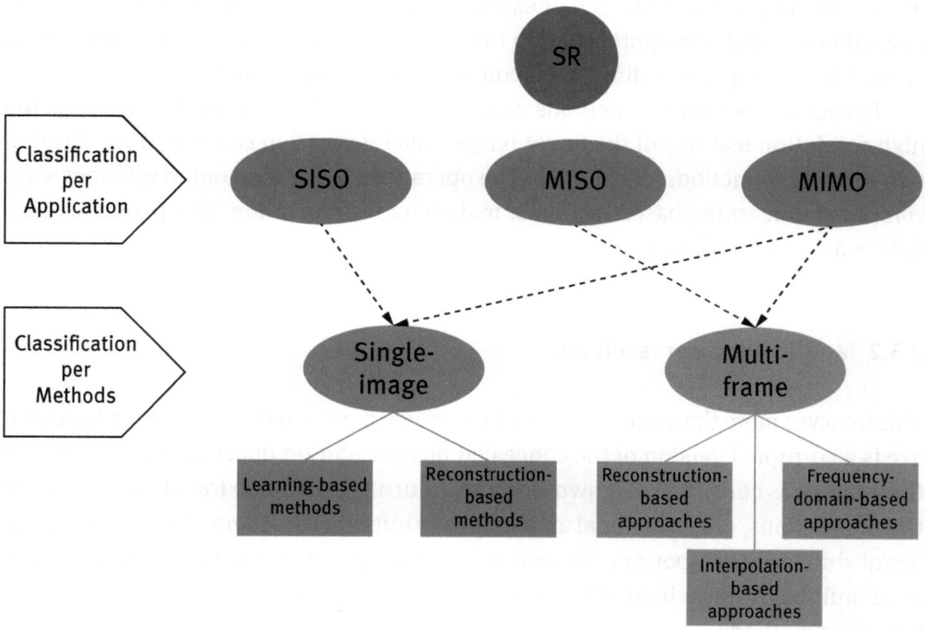

Fig. 3.4: Taxonomy of image super resolution techniques

Historically, methods like maximum likelihood, maximum a posteriori, joint map restoration and Bayesian treatments have been used for super resolution, which involves directly altering the pixels of an image to produce a better image. These methods can also be categorized as spatial domain methods for image enhancement [10].

Generative models find their application here as well by taking the original image and increasing their resolution by simultaneously training two networks that use game theory to adversarially train themselves and produce true-to-life images [7].

3.3.1 Single-image super resolution

Single-image super resolution (SISR) deals with estimating high-resolution images. Given a low-resolution image as an example, the set-up for the original image is not available. SISR techniques can be related to traditional interpolation methods like cubic splines, bicubic and linear interpolation. The job can be simplified to compute the lacking pixel intensities in the high-resolution grid. This can be performed by doing averages of known pixels, which is suitable for smooth parts but needs elaboration when encountering discontinuities, such as corners and edges, since this might lead to blurring of artefacts. There are two implicit types of SISR methods, learning-based methods and reconstruction-based methods.

Learning-based methods involve machine learning approaches to sectionally estimate the high-resolution features of the output image. They may be of various types, pixel based [11], patch based [12] or example based. The example-based methods are those that use multiple approaches on the same image to project internal similarities, approaches like sparse coding, neighbour embedding and so forth.

Reconstruction-based methods require categorical information regarding the high-resolution features of the target image. The information can vary from distribution to energy function, for example. The operations are carried out in different ways compared to learning-based methods, mainly by regularization, sharpening of edge details and deconvolution.

3.3.2 Multiframe super resolution

This involves more than one low-resolution image, during experiments each input image is a corrupted version of the concealed high-resolution image. It is assumed that the image was corrupted by downsampling, blurring and affine transformations. For transformations, only fractional and subpixel shifts are considered to be of value; integral shifts are unimportant. There are three types of super resolution methods that have multiple frames: interpolation methods, frequency domain methods and regularization methods.

The interpolation methods involve deblurring, registration and interpolation. These three steps are conducted using various heuristics. The frequency domain methods aggregate different metrics of high resolution images from the transformations, DCT (discrete cosine transform), DFT (discrete fourier transform), DWT (discrete wavelet transform) or other frequency domain representations of low-resolution images as input. Among all the localized methods, the DWT domain is superior to the others. The regularization methods are only convenient in the case of a limited number of low-resolution frames. The strategies to incorporate some prior features of the unknown high-resolution image are mainly deterministic and stochastic.

3.4 Image upsampling

Image upsampling refers to the task of enlarging an image with an integral value such that the image can fit the new dimensions. Image upsampling usually replicates rows and columns according to the needs of the dimension being scaled to. Deep learning allows an image to be upsampled with better quality along with higher dimensions as well. Image upsampling is the linear opposite of image downsampling, where the image data are reduced to fit the new resolution. Hence, images can be upsampled by either adding new information to them or by using the existing information to produce a larger version of the same image.

Deep learning can be used in this domain of image processing as many algorithms can be used to generate new images from previously known image data sets. This allows neural-network-based models to produce upsampled images from a data set whose images need to be upsampled. Since the algorithm can provide much more accurate upsampled data, this section discusses the various deep learning methodologies that can produce comparable and even better results than previously known techniques.

Traditional image upsampling techniques involve Fourier upsampling, which uses Fourier transformations to produce upsampled images by adding pixel data for the missing values by transforming the k-nearest neighbours (k-NN) of a pixel through a Fourier transform and adding those values. This method usually involves a 4× Fourier transform with reconstruction or alias [13].

A tried and tested deep learning algorithm involves yet again generative adversarial networks called Laplacian Pyramid Generative Adversarial Networks [14] that use a Laplcaian Pyramid to create missing pixel values and uses adversarial training on a generator and a discriminator to form the pyramid itself.

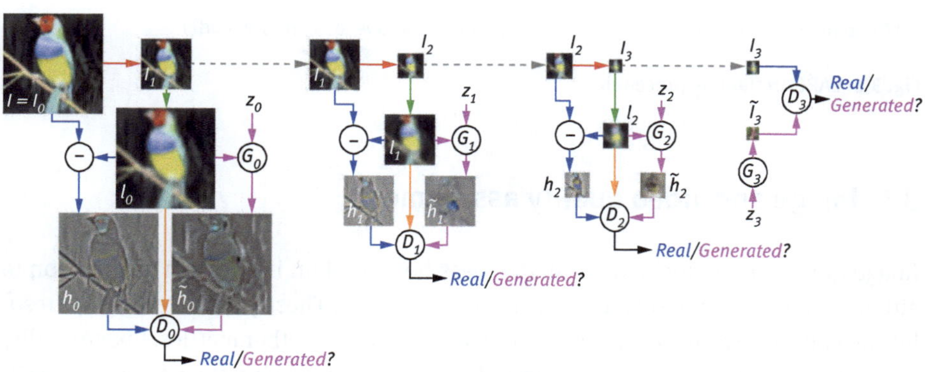

Fig. 3.5: LAPGAN training procedure

3.4.1 Upsampling using neural networks

Deep neural networks are also used in image upsampling. The deep learning method used is known as a deep network cascade [15]. The upscaling is gradual, layer by layer with the introduction of a scaling factor. It eventually attains the higher resolution required. The small refinements are based on searching for non-local self similarities in order to improve the higher-frequency details of the segments to which the image is partitioned. The segments are then absorbed into each layer of the cascading multi-stack network of a local auto-encoder for noise suppression and to enhance the coherence among the overlapping patches enhanced by the aforementioned procedure (Figure 3.6).

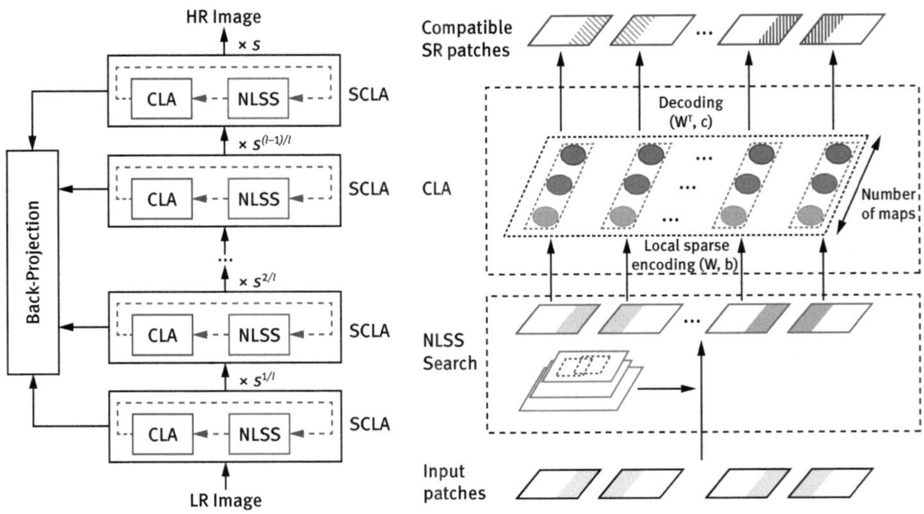

(a) Deep network cascade

(b) Collaborative local auto-encoder

Fig. 3.6: LAPGAN training procedure

3.5 Image and video quality assessment

Image quality assessment is a way to assess how good an image is in comparison to either a reference image or a general set of parameters. These parameters can be resolution, contrast, saturation levels, colour accuracy and or other metrics. Image quality assessment is a necessary component of any image processing system as it makes it possible to compare how the processing application is working.

Traditionally, image quality assessment was done using the visibility of errors (differences) between a distorted image and a reference image using a variety of known properties of the human visual system [16]. Another method used for such quality assessment is called a structural similarity index that extracts the base structure of an image by eliminating the overlying luminance and contrast features and satisfies a few conditions to provide an image that is structurally similar to the original one [17].

More recent deep learning approaches to quality assessment use a CNN to learn patch features in the spatial domain that are used to determine the quality of an image without any reference images for a base comparison [18].

3.5.1 Error sensitivity–based method

According to Wang et al. [19], image quality measurement is used for three primary reasons: to monitor quality for quality systems, to benchmark various algorithms that generate or modify images, and to optimize the image generation process of algorithms or techniques.

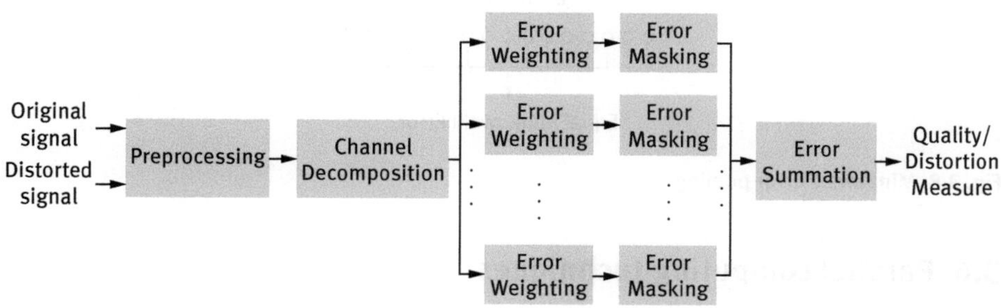

Fig. 3.7: Image quality assessment based on error sensitivity

One image quality assessment technique that is used is the error sensitivity method that uses a full-reference model to compare images and calculate the error between images. First, two images are subjected to different preprocessing techniques, which then decompose them into difference channels. The error for each channel, weighted by a contrast sensitivity function (CSF), is then transferred to an error-masking function, which then sums the error using a summation function, which tells us the quality or the distortion measurement of the image. This summation function can also use a Minkowski error-pooling function. This method has certain drawbacks due to the following assumptions, mainly:

- The reference image is a perfect image
- The visual channels in the images can be simulated by channel transformations
- In each channel, the interactions between coefficients is small enough to be ignored
- The image quality perceived is done by the early vision system
- Active visual processes, such as change of fixation point, are ineffective

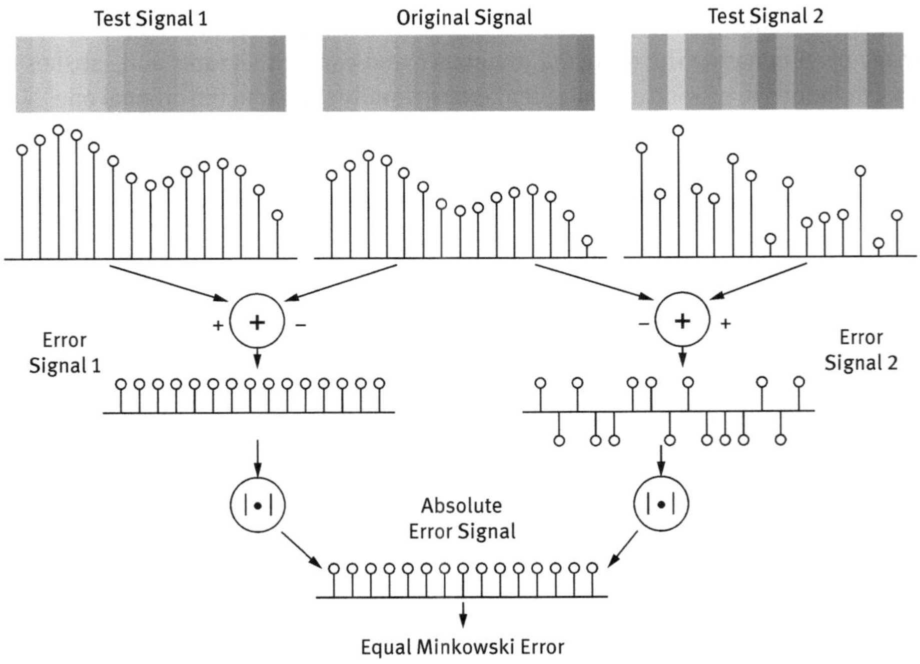

Fig. 3.8: Minkowski error pooling

3.6 Parallel computing techniques

Parallel computing refers to a computing paradigm that uses multiple simultaneously processed threads that can be used for varied applications such as graphic processing, neural network training and hyperthreading. This new methodology has been become famous due to its application in research, with the most common form of hardware used in such systems being multicore processors and graphics processing units. This work relies heavily on the latter for network training. It also looks at the various motivations, shifts and tensions that have surrounded the use of multiple cores instead of one for macro and niche processing in various fields of study [20].

The image processing problems that require vast amounts of time for processing or handling large amounts of data can be addressed with parallel computing techniques. The principal idea involves dividing a problem into simple tasks and solving them concurrently. The best case scenario is attained when the total time can be divided among the total tasks. Not all problems can be solved by parallel image processing, and not all methodologies require coding in a parallel form. Every parallel coded technique has features of an efficiency or a correcting operation. These features mainly include granularity, synchronization, latency, scalability and a defined type of parallel processing.

Granularity determines the number of key units in a frame; it is represented by two types, coarse grained and fine grained. Coarse grained is indexed to having few tasks but each with intensive computing methods, whereas fine grained is indexed to having a large number of small tasks with no specific intensive task, so with less intensive computing.

Synchronization is embedded to ensure that no two processes overlap. Latency is defined as the time of transition of information from request to declaration. Scalability is the metric that indicates the ability of an algorithm to maintain its efficiency by increasing the number of processors in proportion to the size of a problem.

3.6.1 Data parallelism

The objective of any learning algorithm is to minimize the loss function in a methodical manner. In the case of deep neural networks, there is a loss function introduced that is an independent and differential function of the model hyper-parameters. In the case of parallel image processing, methods are generated to iteratively update the parameters with the aim of reducing the net-loss function. One way to do this is to perform mini-batch gradient descent, but it is computationally expensive and time consuming. Hence the weight update procedure needs to be parallelized to achieve accelerated learning.

Data parallelism is a procedure to distribute computing among different machines in a way that computation is performed locally on each machine using the complete model but only for part of the full data set. A representation of data parallelism for updating parameters is demonstrated in Figure 3.9. The most commonly used version of data parallelism is the implementation of parallel stochastic gradient descent (ParallelSGD) [21]. This acts as a substitute for a multibatch stochastic gradient descent algorithm.

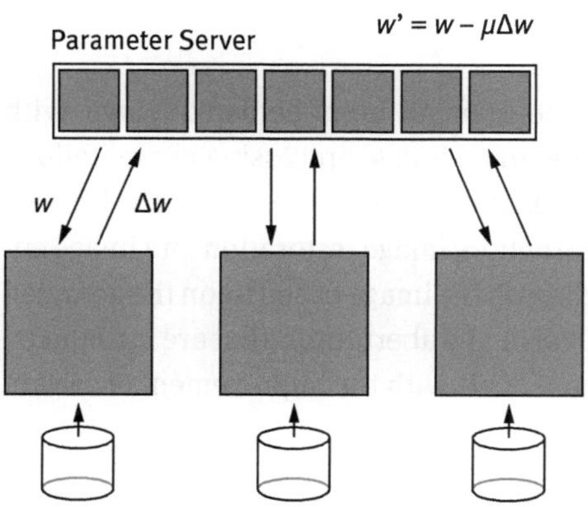

Fig. 3.9: Data parallelism for updating parameters

Once every machine has the data required to go over the distribution of the whole data set, ParallelSGD is run on each machine locally. The communication time and communication cost are the factors that additionally contribute to the net error. Upon updating the parameters, a global communication procedure is performed to send the information to the driver machine, where it will be further be analysed independently. BitTorrent aggregate communication is the protocol that has achieved the best result thus far.

3.7 Modern deep neural architectures

The most widely used application of image processing techniques is in the implementation of deep neural networks. In the contemporary world, almost all types of computational processing can be optimized by the implementation of deep neural architectures [22].

Deep learning techniques were initially applied for adapting to parallel strategies for time series learning [23]. Along time, the computational load for time series learning was multithreaded. Deep neural networks have been applied to several classification and recognition problems on images [24]. Part of its success is due to the automatic feature extraction at different levels of abstraction [25]. In the past decade, there have been multiple designs and heuristics to implement learning in deep neural networks for images. Those that are trending and heavily discussed in the community include deep CNNs (DCGANs) [26], adversarial learning [27], deep residual learning [28] and shallow learning algorithms [29]. We would like to elaborate on the neural architectures created to tackle the different tasks in the chain of image processing. The following subsections discuss the degrees of success attained by deep neural networks in each task for which they have been used.

3.7.1 Image preprocessing

Preprocessing entails any operation performed upon an input-actuated image, with the output being a full image. It is the first step in an image processing chain, following image acquisition. The operations performed during preprocessing can be broadly classified into three categories: image reconstruction, image restoration and image enhancement. Reconstruction consists in building up the image based upon the actuated measurements. Restoration entails the removal of any aberrations that are introduced by the sensor, for instance noise. Enhancement deals with the improvement of certain accentuated features.

3.7.1.1 Image restoration

Neural networks are widely used in image restoration due to their functionality of excellently adapting to the local nature of a problem. They can be implemented for various use cases, usually alternating between restoration and learning cycles. Contemporary neural network processing methods have led to the creation of efficient very-large-scale integration architectures for image restoration, mainly due to their high parallel image processing features.

Dianhui Wang et al. [30] introduced a pattern-learning-based image restoration network for highly deteriorated images. The regularization function that was used to recover the artefacts uses the edge information obtained from the source images. Image similarity was the criterion used by this network to attain, though the judgment is subjective, a quality enhancement. This paper established a mathematical foundation for the use of modern neural networks to perform the task of preserving edge information.

Deng Zhang et al. [31] introduced a neural network designed to eliminate pulse noise. The method was closely based upon anisotropic diffusion methods. The network was termed a pulse-coupled neural network (PCNN) and implemented a denoising method using convolution. The researchers further compared their performance to the traditional filters such as median, Weiner filter, Lee filter and anisotropic diffusion filter. This paper established the benchmark in noise suppression along with edge detail preservation.

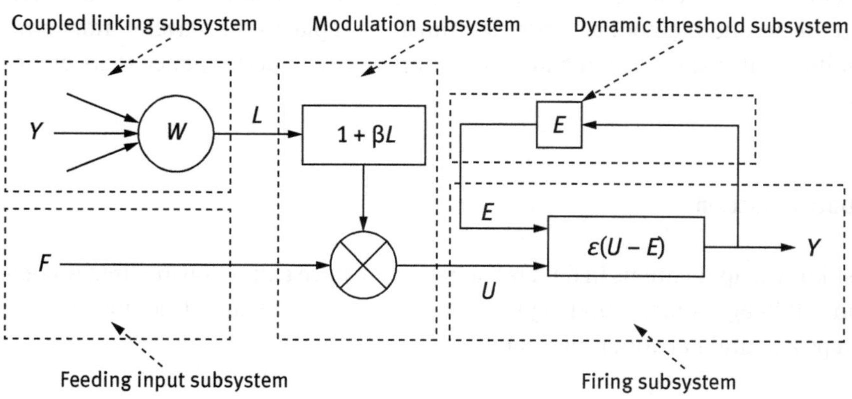

Fig. 3.10: Pulse-coupled neural network model

3.7.1.2 Image reconstruction

Using knowledge from the degrading factors and building up the original image is defined as image reconstruction. Contemporary image restoration techniques have one common deficiency, poor convergence properties. The local minima that the heuristics converge upon are speculative for real imaging applications. The key problem is the recovery of the information of the lost details in the present blurred image. Multilayer Feedforward Neural Network Based on Multi-Valued Neurons (MLMVN), CNN and backpropagation neural network (BPNN) are used to attain high-quality restored images by fast neural computation. They are commonly used because they do not employ lengthy training algorithms.

As a recent trend, Suhasini et al. [32] showcases the use of a BPNN for blur parameter identification. The BPN is trained upon the parameters extracted from a degraded image and the network is simulated to restore the image. It employs deblurring with convolution with a Lucy Richardson algorithm. Backpropagation neural networks have displayed great success in areas requiring high computation rates.

3.7.1.3 Image enhancement

Image enhancement involves sharpening image features like boundaries, contrast and edges. There is no increment in inherent information content in the image data. The dynamic range of these features is increased so they can be more easily visually detected. The prime objectives are noise reduction, removal of inhomogeneities in the background and feature extraction.

Entezarmahdi et al. [33] set the benchmark in neural image enhancement methods compared to stationary methods. This paper defines two transform-based super resolution methods for image enhancement. As per the first method, the wavelet transform coefficients of the lower-resolution image is used as data to train a neural network. In the second method, a contourlet transform replaces the wavelet transform. The results of this experiment demonstrated superiority in performance to its predecessors.

3.7.2 Data reduction

The best-known applications in data reduction are feature extraction and image compression. All image compression algorithms involve encoding and decoding, and for these steps a neural network is designed.

3.7.2.1 Image compression

There are two widely observed compression techniques based on modular approaches and on neural networks. The major advantages of using an artificial neural network (ANN) is that the model is adaptable, so when trained on any particular image data set, it gives significantly better compression rates.

Jin Wang et al. [34] use a custom-designed neural network based upon wavelet transforms. Their use case was ECG data compression, but their novel approach is to continuously adjust the initial weights while accelerating the convergence speed.

They compare the performance of their network to a wavelet compression algorithm, their results were more satisfactory compared to BPNN and principal component analysis (PCA) methods.

3.7.2.2 Feature extraction

Feature extraction can be seen as a special form of data reduction in which the aim is to find the subset of variables that are informative based on the image data. It is crucial to make a distinction between the involvement of supervised and unsupervised ANNs for feature extraction. Both learning methods have established significant benchmarks in feature extraction compared to traditional techniques such as PCA.

Wu Jiang et al. [35] have developed a novel algorithm for image-based feature extraction built upon the PCNN mentioned earlier. Execution time and size are the parameters used to evaluate the results. The experimental results proved that they outperformed traditional Euclidean algorithms.

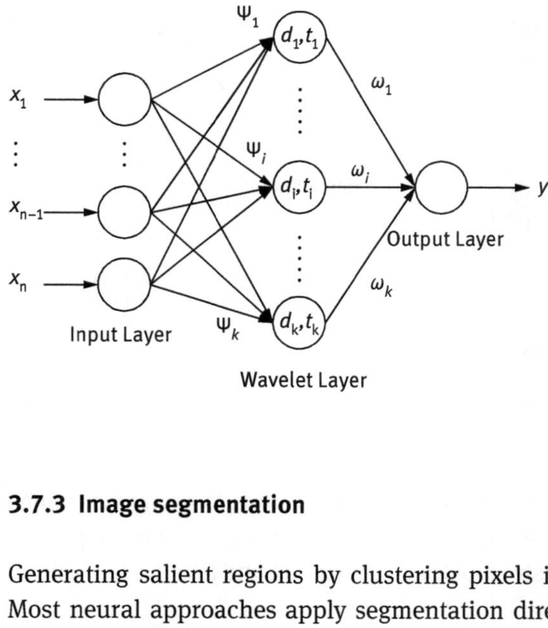

Fig. 3.11: Wavelet-based neural network model

3.7.3 Image segmentation

Generating salient regions by clustering pixels is the aim of image segmentation. Most neural approaches apply segmentation directly to pixel data, whereas others approach it by generating patterns at the feature level. Hence the two approaches can be simplified to clustering and pattern recognition.

3.7.3.1 Clustering

All images contain well-formed information that can be inferred by humans upon viewing the images. When approaching learning methods in an unsupervised method and dealing with the congregation of unlabeled data, the process of clustering may be considered. Clustering basically involves determining the spatial thresholds that reveal the similarity and dissimilarity of certain objects in connection with others.

Li Jing et al. [36] propose a method based upon general regression neural network (GRNN) and fuzzy c-means (FCM) clustering algorithms. Their method involves the creation of pattern neurons to embed and extract identified locations based upon the embedding strength on the human visual system (HVS). PSNR (Peak Signal to Noise Ratio) and BCR (Bit Correction Rate) were used to evaluate the performance of their method.

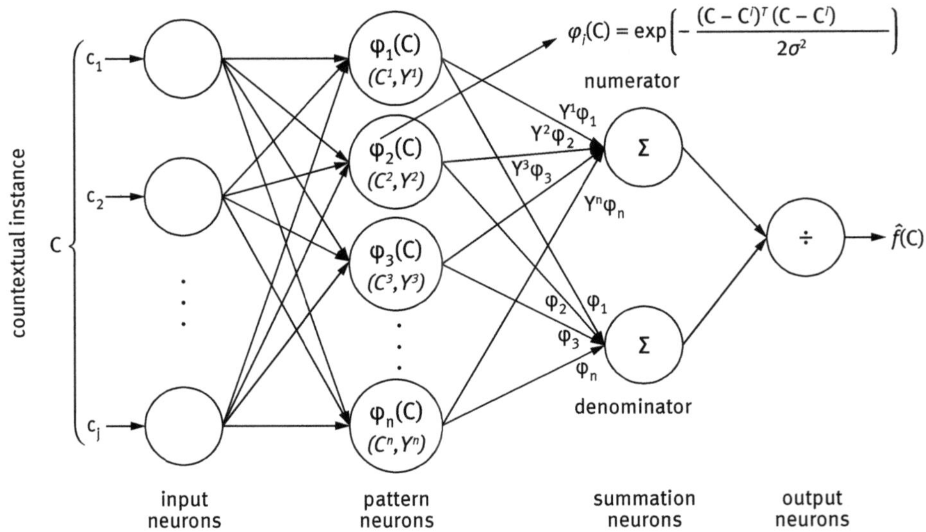

Fig. 3.12: General regression-based neural network model

3.7.3.2 Pattern recognition

The perceptive powers of the human mind allows us, over time, to recognize trained objects. Pattern recognition is defined as action based upon the category of pattern identification. An ANN is an impressive tool for classification and pattern-recognition tasks.

Budiwati et al. [37] describe one of the most popular neural networks, which was designed for Japanese character recognition and based upon a BPNN. There are multiple writing systems in the Japanese language, and there was a need for pattern recognition to distinguish the writing styles of Kanji and Kana. The performance of the neural network was compared to the computational receiver operating characteristic (ROC). It set a benchmark in pattern recognition as it attained an accuracy of 92.29% from 3,956 images to train upon.

3.7.4 Object recognition

Object recognition consists of spatially finding the orientation and position of instances of objects in an image. The most common purpose of object recognition is to assign a class label to a recognized object. Neural networks can be trained to locate individual objects based upon the data embedded within the pixels. Broadly speaking, there are two types of object recognition methods, template matching and feature-based matching.

3.7.4.1 Feature-based matching

Based upon the features of an object, they can be recognized even if they are partially obstructed from view. The neural approaches based upon feature-based object recognition involve a Hopfield ANN, fuzzy ANN and traditional feedforward ANN.

In 2010, a face recognition method was introduced by Jong-Min Kim et al. [38] that performed intelligent classifications by incorporating a multilevel neural network (MLNN) along with PCA simultaneously. Each image decides the initial set of weights, but the feature vector reduces the dimensions of the image at the same time that facial recognition is performed.

3.7.4.2 Template matching

Template matching involves the action of discovering small parts of an image and matching it to a template image. Due to the weight-sharing restrictions observed on local connections in a CNN, they provide an efficient way to constrict the complexity of feedforward neural networks. Generally layers are trained over recognizable features in sequence.

Yilmaz et al. [39] provide a hybrid approach that includes histogram equalization, thresholding and grey-level transformations in license plate recognition. Their algorithm sets a benchmark in the domain of template matching with a very robust success rate. It first extracts features, learns upon them and further uses them to extract characters and eliminate noise.

3.7.5 Image understanding

Image understanding is the most problematic area of image processing. It entails building correlations with the content of an image using techniques such as segmentation and recognition. It involves two major steps, scene analysis and object arrangement.

3.7.5.1 Scene analysis

Scene analysis is quite different in comparison to the recognition of different objects. It involves inherent changes in variation, illumination and noise. Neural networks act as classifiers by generating multiple weighted connections.

Seiji et al. [40] set a benchmark in image searching by introducing a novel architecture, a sandglass-type neural network. It was built for the use case of browsing searches and similarity-based image retrieval. The authors use the maxi-min distance algorithm (MDA) for image preprocessing.

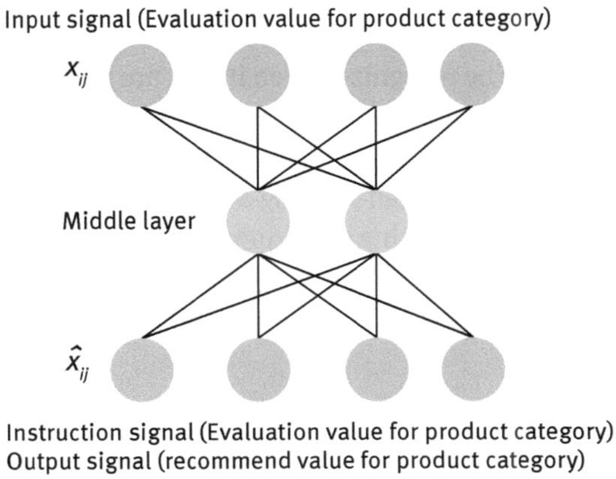

Fig. 3.13: Sandglass-type neural network model

3.7.5.2 Object arrangement

Any fragment of an image that could be categorically recognized as a single unit is referred to as an object. The most popular problem statement tackled using object arrangement is that of real time object tracking. Neural networks are most commonly used for tracking and arrangement purposes because of their capability of non-linear mapping among varied complexities.

Hiremath and Kodge [41] have established modern benchmarks in locating object boundaries using active contours. They employ a *k*-NN along with neural classifiers. The performance criteria of MSE (Mean Square Error) and SD (Standard Deviation) is used for sampling and distribution of these features.

3.7.6 Image optimization

Image optimization deals with the selective process of choosing the best element from a given set of elements based upon some deciding criteria. Optimization problems include a wide variety of functions, each one catering to a specific domain. The most common image optimization technique is graph matching.

3.7.6.1 Graph matching

Labelling the significant objects in an image by identifying them is defined as image interpretation. Image interpretation conducted across the graphical representation is termed graph matching. The represented graph is further partitioned in accordance with the criteria formulated for segmentation into clusters.

Various neural network architectures are applied subject to a particular graph matching problem. Zinkevich et al. [21] introduced a novel architecture based upon the multilevel perceptron neural network. It was designed for the use case of thumbprint classification, and training is performed by a vanilla ANN. Its performance criterion is the MSE and sets the lowest MSE value at less than 0.1.

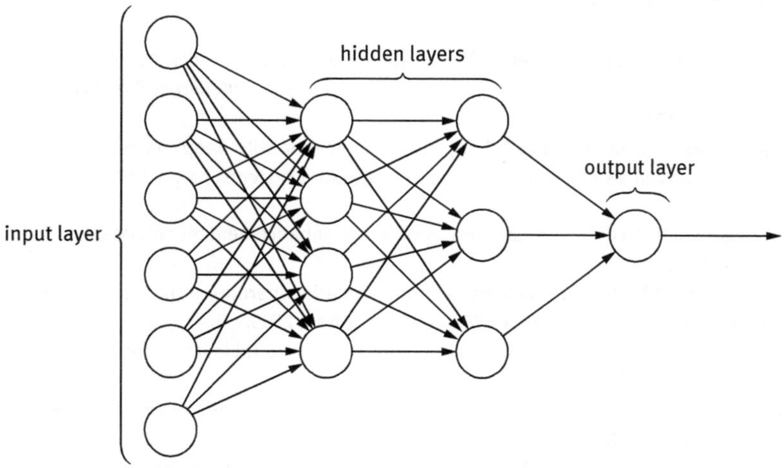

Fig. 3.14: A multilevel perceptron network model

3.8 Conclusion

This chapter presented an in-depth explanation of various deep learning techniques that are used in various tasks performed along the image processing chain. Huge amounts of neural architectures have been developed by researchers in recent decades, and we provided an explanation regarding techniques implemented for particular tasks in image processing for each model. Case studies have shown that hybrid models demonstrate superior performance in comparison to conventional models. The model that sets the highest benchmark was described briefly, and each task of image processing used in the model was described in detail.

Even though conventional models have high accuracy, newly defined hybrid models are tailored to a particular use case. Combining these conventional models in accordance with their use case yields higher accuracy and contributes to the individual performance of the models. Hence, the highly iterative and learning-based approach of deep learning techniques shows better performance for tasks in image processing.

Bibliography

[1] Dechter R. Learning while searching in constraint-satisfaction-problems. In Kehler T, editor, *AAAI*, pages 178–185. Morgan Kaufmann, 1986.

[2] Kelley HJ. Gradient theory of optimal flight paths. *ARS Journal*, 30(10):947–954, 1960.

[3] Igor Aizenberg JPLV Naum N. Aizenberg. *Multi-Valued and Universal Binary Neurons*. Springer US, 2000.

[4] Hadsell R, Erkan A, Sermanet P, Scoffier M, Muller U, and LeCun Y. Deep belief net learning in a long-range vision system for autonomous off-road driving. In *Proc. Intelligent Robots and Systems (IROS'08)*, 2008.

[5] Krizhevsky A, Sutskever I, and Hinton GE. Imagenet classification with deep convolutional neural networks. In *Advances in neural information processing systems*, pages 1097–1105, 2012.

[6] Maini R and Aggarwal H. A comprehensive review of image enhancement techniques. *CoRR*, abs/1003.4053, 2010.

[7] Radford A, Metz L, and Chintala S. Unsupervised representation learning with deep convolutional generative adversarial networks. *CoRR*, abs/1511.06434, 2015.

[8] Gharbi M, Chen J, Barron JT, Hasinoff SW, and Durand F. Deep bilateral learning for real-time image enhancement. *CoRR*, abs/1707.02880, 2017.

[9] Yang J, Ren P, Chen D, Wen F, Li H, and Hua G. Neural aggregation network for video face recognition. *CoRR*, abs/1603.05474, 2016.

[10] Yang J and Huang T. Image super-resolution: Historical overview and future challenges. *Super-resolution imaging*, pages 20–34, 2010.

[11] Liu P and Fang R. Learning pixel-distribution prior with wider convolution for image denoising. *CoRR*, abs/1707.09135, 2017.

[12] Figueiredo MAT. Synthesis versus analysis in patch-based image priors. *CoRR*, abs/1702.06085, 2017.

[13] Shan Q, Li Z, Jia J, and Tang CK. Fast image/video upsampling. *ACM Transactions on Graphics (TOG)*, 27(5):153, 2008.

[14] Denton EL, Chintala S, Fergus R, et al. Deep generative image models using a laplacian pyramid of adversarial networks. In *Advances in neural information processing systems*, pages 1486–1494, 2015.

[15] Schlemper J, Caballero J, Hajnal JV, Price AN, and Rueckert D. A deep cascade of convolutional neural networks for dynamic MR image reconstruction. *CoRR*, abs/1704.02422, 2017.

[16] Mannos J and Sakrison D. The effects of a visual fidelity criterion of the encoding of images. *IEEE transactions on Information Theory*, 20(4):525–536, 1974.

[17] Wang Z, Bovik AC, Sheikh HR, and Simoncelli EP. Image quality assessment: from error visibility to structural similarity. *IEEE transactions on image processing*, 13(4):600–612, 2004.

[18] Kang L, Ye P, Li Y, and Doermann D. Convolutional neural networks for no-reference image quality assessment. In *Proceedings of the IEEE conference on computer vision and pattern recognition*, pages 1733–1740, 2014.

[19] Wang Z, Bovik AC, and Lu L. Why is image quality assessment so difficult? In *Acoustics, Speech, and Signal Processing (ICASSP), 2002 IEEE International Conference on*, volume 4, pages IV–3313. IEEE, 2002.

[20] Asanovic K, Bodik R, Demmel J, Keaveny T, Keutzer K, Kubiatowicz J, Morgan N, Patterson D, Sen K, Wawrzynek J, Wessel D, and Yelick K. A view of the parallel computing landscape. *Commun. ACM*, 52(10):56–67, October 2009.

[21] Zinkevich M, Weimer M, Li L, and Smola AJ. Parallelized stochastic gradient descent. In Lafferty JD, Williams CKI, Shawe-Taylor J, Zemel RS, and Culotta A, editors, *Advances in Neural Information Processing Systems 23*, pages 2595–2603. Curran Associates, Inc., 2010.

[22] Blum C, Puchinger J, Raidl GR, and Roli A. Hybrid metaheuristics in combinatorial optimization: A survey. *Applied Soft Computing*, 11(6):4135 – 4151, 2011.

[23] Coelho IM, Coelho VN, da S. Luz EJ, Ochi LS, Guimarães FG, and Rios E. A gpu deep learning metaheuristic based model for time series forecasting. *Applied Energy*, 201:412 – 418, 2017.

[24] Sun Y, Chen Y, Wang X, and Tang X. Deep learning face representation by joint identification-verification. In *Advances in neural information processing systems*, pages 1988–1996, 2014.

[25] Ojha VK, Abraham A, and Snášel V. Metaheuristic design of feedforward neural networks: A review of two decades of research. *Engineering Applications of Artificial Intelligence*, 60:97 – 116, 2017.

[26] Krizhevsky A, Sutskever I, and Hinton GE. Imagenet classification with deep convolutional neural networks. In *Advances in neural information processing systems*, pages 1097–1105, 2012.

[27] Lowd D and Meek C. Adversarial learning. In *Proceedings of the Eleventh ACM SIGKDD International Conference on Knowledge Discovery in Data Mining*, KDD '05, pages 641–647, New York, NY, USA, 2005. ACM.

[28] He K, Zhang X, Ren S, and Sun J. Deep residual learning for image recognition. In *The IEEE Conference on Computer Vision and Pattern Recognition (CVPR)*, June 2016.

[29] Dechter R. Enhancement schemes for constraint processing: Backjumping, learning, and cutset decomposition. *Artificial Intelligence*, 41(3):273 – 312, 1990.

[30] Man Z, Lee K, Wang D, Cao Z, and Khoo S. Robust single-hidden layer feedforward network-based pattern classifier. *IEEE transactions on neural networks and learning systems*, 23(12):1974–1986, 2012.

[31] Zhang D, Mabu S, and Hirasawa K. Image denoising using pulse coupled neural network with an adaptive pareto genetic algorithm. *IEEJ Transactions on Electrical and Electronic Engineering*, 6(5):474–482, 2011.

[32] Suhasini A, Palanivel S, and Ramalingam V. Multimodel decision support system for psychiatry problem. *Expert Systems with Applications*, 38(5):4990–4997, 2011.

[33] Entezarmahdi SM and Yazdi M. Stationary image resolution enhancement on the basis of contourlet and wavelet transforms by means of the artificial neural network. In *Machine Vision and Image Processing (MVIP), 2010 6th Iranian*, pages 1–7. IEEE, 2010.

[34] Wang J, She M, Nahavandi S, and Kouzani A. Human identification from ecg signals via sparse representation of local segments. *IEEE Signal Processing Letters*, 20(10):937–940, 2013.

[35] Jiang H, Wang J, Yuan Z, Wu Y, Zheng N, and Li S. Salient object detection: A discriminative regional feature integration approach. In *Proceedings of the IEEE conference on computer vision and pattern recognition*, pages 2083–2090, 2013.

[36] Cliff MA, Dever MC, Hall JW, and Girard B. Development and evaluation of multiple regression models for prediction of sweet cherry liking. *Food Research International*, 28(6):583–589, 1995.

[37] Budiwati SD, Haryatno J, and Dharma EM. Japanese character (kana) pattern recognition application using neural network. In *Electrical Engineering and Informatics (ICEEI), 2011 International Conference on*, pages 1–6. IEEE, 2011.
[38] Kim JM and Kang MA. A study of face recognition using the pca and error back-propagation. In *Intelligent Human-Machine Systems and Cybernetics (IHMSC), 2010 2nd International Conference on*, volume 2, pages 241–244. IEEE, 2010.
[39] Yilmaz A, Javed O, and Shah M. Object tracking: A survey. *Acm computing surveys (CSUR)*, 38(4):13, 2006.
[40] Ito S, Mitsukura Y, Fukumi M, Akamatsu N, and Omatu S. *Scene image analysis by using the sandglass-type neural network with a factor analysis*, volume 2, pages 994–997. Institute of Electrical and Electronics Engineers Inc., 2003.
[41] Kodge BG and Hiremath PS. Elevation contour analysis and water body extraction for finding water scarcity locations using DEM. *CoRR*, abs/1404.3002, 2014.

Abantika Choudhury* and Moumita Deb

4 Connecting cities using smart transportation: an overview

Abstract: This chapter focuses on connected cities in terms of smart transportation. A connected city controls available resources in such a way that it can efficiently improve the economy and societal outcomes. Many data are generated from people, systems and things in cities. Thus, data generated from various resources are considered to be the most scalable asset of a connected city. Heterogeneous data are difficult to organize, interpret and analyze. Generally the data generated from various sources are very large and heterogeneous as well because they are generated from heterogeneous environments like water, traffic, energy and buildings. Different contexts, such as multidimensional ones, for example databases, data mining, artificial intelligence and distributed systems communities, are good areas that may be equipped to deal with the challenges of big data in connected cities.

Keywords: Connected city, VANET, RSU, OBU, Big data, IoT

4.1 Introduction

Smart cities aim to develop urban areas by integrating Internet of Things (IoT) technology and information and communication technology (ICT) to handle the various resources of cities. A smart city is a combination of various systems that are interconnected with each other. Systems like health care, transportation, entertainment, public services and employment are involved in smart city systems. A connected city combines technology, government and society to enable different features like, for example, a smart environment, smart economy, smart living and smart mobility. Nowadays, there is a strong need for intelligent environments that allow for high-quality lifestyles. Smart transportation, smart communications and smart networks are emerging to facilitate such lifestyles. Connected cities require huge resources from authorities and efficient planning. One important aspect of a connected city is a connected network of transportation that can communicate with other types of systems to facilitate smart communication among vehicles. The dynamic nature of a connected city makes it different from other cities with respect to its intelligent infrastructure. Ongoing research approaches focus on developing smart cities using smart communication and transportation. In this chapter, a recent trend in smart transportation is described along with its merits and demerits.

***Corresponding author: Abantika Choudhury,** RCC Institute of Information Technology
Moumita Deb, RCC Institute of Information Technology

https://doi.org/10.1515/9783110551433-004

A connected city uses technology to improve the quality of services. Rick Robinson, Director of Technology for Amey, UK, proposed that the architecture for a smart city (Figure 4.1) consists of six layers: goals, people, ecosystem, soft infrastructures, city systems and hard infrastructures. A connected city should possess a set of goals according to the requirements of that locality. Such goals of a smart city can be achieved by transportation, education, health, economy and so forth. Robinson also asserts that the connected city concept can be used to systematically upgrade a city and the utility of designing infrastructures like a smart city is to enable the innovation of communities within a city. Community is a vital element of a city's ecosystem. Communities provide support to the infrastructure of a smart city. The architecture shows that a variety of soft infrastructures (e.g., leadership, innovation forums and network and community organizations) play an important role in smart cities. The diagram in Figure 4.2 also shows the hard infrastructure. The challenge in designing a smart city is to build the proper infrastructures to provide services for communities and society. These systems feed, transport, educate and provide health care for residents, supporting communities and businesses. To design a smart city, it is necessary to remove constraints from existing cities. The infrastructures and operations of a city are a challenge for service providers. Those constraints serve to degrade the services of a smart city. The people, communities and organizations of a city should understand the necessity, opportunities, technology and challenges of building a connected city.

4.2 Components of smart cities

Intelligent infrastructure is a combination of software and hardware systems. It is also combined with various network communication and sensor nodes. We mainly focus our chapter on smart transportation for connected cities. A classification of connected cities is given in Figure 4.1. A connected city is operated by Internet connection. A city having broadband Internet facilities is called a broadband city. The Ubiquitous City (U-City) is a city where a substructure of ubiquitous computing is applied for different functionalities of urban systems. Wireless cities are cities where wireless access is available to support outdoor hot spots that provide public Internet access in several areas, for example parks, bus stops and restaurants. Digital city refers to an online network that operates locally and delivers local information like about some event or entertainment, a visitor's guide or e-commerce. An eco-city is designed to promote a healthy environment.

Intelligent transportation systems (ITSs) provide modern services for traffic management and vehicular transportation in urban areas. ITSs make transport safer and smarter. Technologically, an ITS is different from traditional transportation systems. It provides more advanced applications. ITS systems provide predictive techniques to compare with historical data and design the enriched model. Smart transportation deals with smart vehicles that can communicate over the Internet. It consists of vehicle-to-infrastructure (V2I) and vehicle-to-vehicle (V2V) communication. V2I means communication between on-board units (OBUs) and roadside units (RSUs); V2V means communication between different vehicular OBUs. Vehicular ad hoc networks (VANETs) efficiently support various applications of ITSs. Transport infrastructure is a backbone to any effective smart city. The architecture of a connected city with smart transportation is given in Figure 4.2.

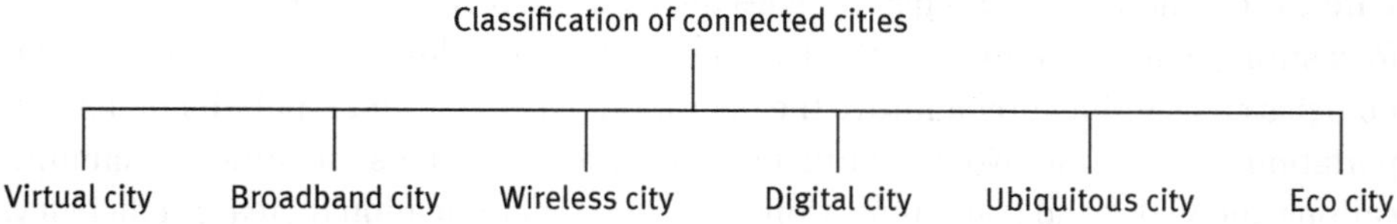

Fig. 4.1: Classification of connected cities

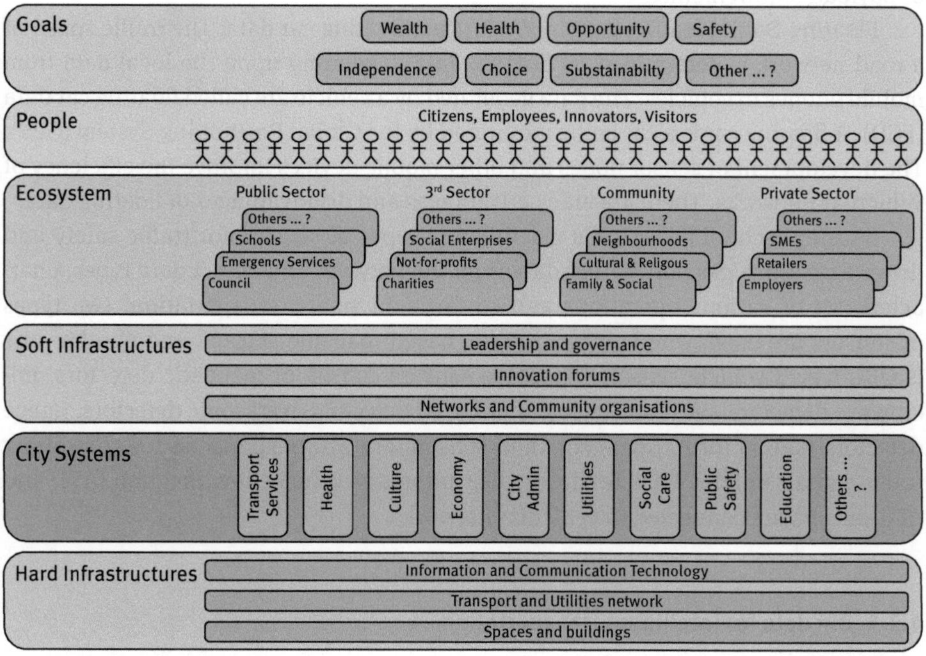

Fig. 4.2: Components of connected/smart city architecture

4.3 Technologies for smart transportation

Different technologies are used for smart transportation. The technologies are as follows:

1. Wireless communication
2. Floating data (regarding vehicles)
3. Sensing technologies

Wireless communication: Since vehicles are continuously in motion, there is a need for a strong wireless communication to interconnect vehicles in an ITS. Wireless technologies can significantly improve the operation, efficiency and reliability of a transportation system. The two main functions of communications are critical communications and wideband communications. Critical communication increases efficiency, safety and reliability between vehicles and infrastructure and wideband communications are used for payload or passenger services [1]. It is necessary to accurately design the physical interface for each special environment, vehicle dynamic and passengers or payload requirements.

Floating Data: Floating data in VANETs are floating car data. The traffic speed of a road network is determined by floating data depending upon the local data from mobile phones to vehicles. The data generated by a vehicle are called floating car data (FCD). A floating car is a car being monitored by the Global Positioning System (GPS). The development of sensor nodes and other mobile devices improve the efficiency of vehicular networks. There are many advantages and disadvantages of floating data.

Sensing technologies: RSUs use different types of sensors for traffic safety and driver assistance. Sensors gather data from the network and depict data types, characteristics of various operations and so forth. In smart transportation, two types of sensors are used. One is infrastructure-based roadside sensors and another one is OBU-based vehicle sensors. Roadside sensors consist of magnetic detectors, microwave detectors, pressure detectors, infrared detectors, ultrasonic detectors, image detectors and so forth. Some common sensors are also responsible for controlling traffic signals in ITSs. Vehicle sensors like automatic vehicle identification (AVI) and GPS sensors are connected to vehicular networks.

4.3.1 Big data for intelligent transportation

Big data has brought many changes in the field of ITS. Emerging technologies are providing the means for enriching ITS performance. Existing ITS technologies are not capable of handling heterogeneous big data. Therefore, it is necessary to develop new tools and technologies to deal with heterogeneous big data [2].

Intelligent transportation using big data provides opportunities to resolve various transportation problems. Data have been applied for transportation analysis, and many challenges have been raised. Data change with time and place. So there is a need for data collection and cleaning. Several types of data have several advantages and limitations for transportation, and one data type alone is not capable of describing the overall state of a transportation system. Data processing and data mining techniques (e.g., natural language processing and analysis of streaming data) require further modification for effectively utilizing real-time traffic information. In this scenario, a demand for big data to solve practical problems of intelligent transportation arises [3].

Deploying big traffic data on the cloud [4] constitutes cloud computing. A smart city begins with the IoT, which involves big data and cloud computing. Big data are transferred in real time in a streaming process over a communication network for storage. It is combined with an extract, transform and load (ETL) process. A standardized structure format is combined with open-source software to develop planning tools to provide a platform to gather public information efficiently to create apps and maps.

4.4 Challenges of big data for smart cities [5]

Smart cities are very dynamic and evolving in character. Many challenges are associated with developing big data applications to make cities smart. These challenges must be addressed to develop smart applications as much as possible. The challenges are related to, for example, real-time analytics, big data tools, accessibility and cost and directly affect the services and performance of smart city applications [6].

4.4.1 Sources of data and their characteristics

Data collected from many sources come in various formats. Most data are not in a structured format. Such unstructured data can be video, audio or images, for example. A database management system (DBMS) is required to convert data into a structured format [7]. More information like validity, volatility, value, variability [8, 9], is also included in systems. The methodologies and the software tools used in data mining are not able to handle the complexity of large size data. This is also a major challenge. There are other challenges like hidden big data analytics architecture, time evolving data, distributed mining, evaluation, compression and visualization [6] that might arise in the future. Big data utilization in smart city applications causes many difficulties in different areas. Gathering data from multiple sources in different formats is very complicated. Unstructured data are very hard to categorize and organize. Thus it is a challenge to come up with an accessible way to implement on applications.

4.4.2 Data and information sharing

Another challenge is serving data or information in various cities. Each government maintains its own data warehouse for public information. Some are warehouses are prohibited from sharing data with other entities. Challenges aris with respect to data security when gathering or using big data; it is necessary to ensure the privacy of citizens [8]. Applications for smart cities need to include strategies to overcome obstacles as much as possible to achieve seamless information sharing [10]. Data semantics are hard to understand, and extracting information based on real-time data poses a challenge. This makes building a knowledge-based data set for smart cities a formidable task.

4.4.3 Quality of data

Maintaining data quality is another challenge. Data captured from different sources are stored in separate databases and follow no standard formats [11]. Collecting data from public sources and collaborating with many providers affect the data format, which makes the data unstructured and inconsistent among other things. This undermines confidence in the data. A challenge may also arise when outputs are erroneous. Erroneous results are propagated to other users, who may not be aware of such issues.

4.4.4 Security and privacy

Security and privacy are another big challenge facing smart cities. Databases need high-level security to protect various data, such as keeping confidential data related to the government out of the hands of unauthorized users. A high level of security is also needed for smart applications as the data move over various networks, some of which are open or unsecured [8]. The technologies used in big data, like Hadoop and Cassandra, must also also deal with security issues [12]. It is also necessary to respect the privacy rights of organizations to secure data as the data travel through different network layers and are used by the components of smart city applications. Different data, like retail sales history, financial records and medical records, for example, provide personal details about people. Providing a high level of privacy policies surrounding big data poses a big challenge for application developers and users of smart cities.

4.4.5 Cost

Another challenge is the cost incurred by government authorities when they from using ICT. An energy usage reduction system [13] is a new system for government energy usage. The components or features of the system are designed to supervise consumption and record data on usage. To overcome the huge energy consumption problem a smart energy management system was created that is costly to implement. It hampers the smart city. The testing of a traffic signal system of a smart city is very costly in terms of resources and traffic problems [14]. Smart city infrastructure and applications [13] require expensive hardware and software to develop and monitor the system.

4.4.6 Smart city population

The population also represents a major challenge for smart cities. A city's population has a huge impact on the size of data. The size of data generated depends upon the population; if the population grows, then the amount of data generated increases. This huge amount of data creates traffic congestion, data loss, broadcast storms, pollution and increasing social inequality [15]. Applications for smart cities must evolve quickly and efficiently to tackle the different types of big data. The challenges discussed earlier have different impacts and effects on smart applications for smart cities. Many difficulties and complexities arise in this scenario. Achieving real-time responses [16]highly depends on being able to meet these challenges.

4.5 Different systems for constructing smart cities

A lot of research is being conducted on smart/intelligent transport for smart cities. Various kinds of ITS for smart cities are being developed to solve different types of problems regarding traffic and transportation. Based on the current trends in research, ITSs can be classified into four main systems:

1. Advanced traveller information system (ATIS)
2. Advanced traffic management system (ATMS)
3. Advanced public transportation system (APTS)
4. Emergency management system (EMS)

ATIS: This kind of system analyses information to help travellers on their way to their destination. An ATIS system can operate through information or data supplied from vehicles or traffic management centres. An ATIS system can deliver information via cellphone, computer, radio, television and so forth. It provides services like mapping or text messaging to automobiles through an in-vehicle information system obtain from some mass communication or personal communication device. For data dissemination, ATIS needs a large amount of traveller-related data to analyse, process and store. A geographical information system (GIS)-based ATIS [17] is very useful for storage and graphical representation of information.

ATMS: This kind of system is used to help reduce traffic congestion in urban areas to enhance the efficiency of existing infrastructures. An ATMS deals with congestion problems on urban roadways. Problems with traffic congestion are caused by both regular traffic patterns and other unpredictable incidents. The system tries to retrieve data given by RSUs. It also tries to develop strategies to optimally regulate traffic in a single intersection throughout a given area. These systems rely on various types of messaging to provide relevant traffic information.

APTS: This kind of system is used to ensure the safety of public transportation. It applies information technologies and transportation management to public transit systems. With the help of these technologies, it increases the efficiency of various transportation operations. APTS applications include automatic vehicle location systems, bus arrival notification systems, real-time passenger information systems and systems providing priority of passage to buses at signalized intersections. Public transportation systems comprise GISs. A GIS is a DBMS. The databases are sets of location coordinates. A GIS is used to maintain bus schedule information in urban areas on different routes and provide information to travellers. Traveller information in an APTS system includes pretrip information so that travellers can plan their trips accordingly. Pretrip information is provided in terminals or at roadside locations or in vehicles.

Connected and smart cities are unlike earlier models of cities in that they rely on ICT to meet their goals. The model of a smart or connected city is vital because of the evolution of ICT in recent years. The ICT trajectories and other related factors, like urbanization rates, show that connected city concepts will become very important in the near future. The essential thing about a connected city is that it needs an intelligent infrastructure. Transportation in a connected city communicates with various systems. Connected vehicles and connected travellers perform data exchange with various systems like employment, entertainment, health care, residential, utilities, retail and public services. The connected/smart city using smart transportation is a challenging issue nowadays. The challenges like a huge number of vehicles, fast mobility, and rapid dynamic charasteristics of vehicles are required to be overcome by government authority. Intense research on this topic can only improve the scenario.

4.6 Intelligent infrastructure [18]

Traditional infrastructures have evolved rapidly into intelligent infrastructures. Various infrastructures, from homes to bridges, from bridges to pipelines and from pipelines to street lights, have become controllable and are able to interact with various infrastructure elements and humans [19]. The components of intelligent infrastructures are connected to the Internet and they can work together to collect contextual data from multiple sources to carry out different functions [20]. Thus, an intelligent infrastructure is more cost-effective, more useful and more adjustable than a traditional infrastructure. Internet of Things denotes a world where the infrastructure is built to link dynamic intelligence through integrated sensors.

4.7 Process innovations [18]

In a connected city, data are available from an intelligent infrastructure. But it is difficult to interpret the data to create value from them and make intelligent decisions. Two main things in the process that are used in connected cities are big data and crowd sourcing.

4.8 Use of city management platforms [21]

The Smartphone Probe Demonstration Project in Fukuoka in 2011 established the Fukuoka Directive Council, a partnership between industry, academia, government and the private sector, to formulate a growth strategy for the Fukuoka metropolitan area and to investigate smart city policies. The council is made up of five working groups (on tourism, environment, food, human resources and urban development), with Hitachi being mainly involved with the environmental working group, which includes participation in vision-building and demonstration projects aimed at making the Fukuoka metropolitan area a leader in smart mobility.

In fiscal year 2013, the environmental working group demonstrate established a project for using smartphone probes to obtain information on “people flow”. Smartphones were pre-installed with an application that collected GPS and accelerometer data. With the cooperation of the Fukuoka City Transportation Bureau, the project also collected data from smart cards when users got on and off the subway. The data were obtained after the trial period had expired. To visualize information on people’s movements, the route travelled and means of transportation were determined automatically from the smartphone GPS and accelerometer data. In the case of subway rides where GPS signals cannot be received, the information was obtained from smart card trip records. The demonstration project automatically classified trips into four categories (walking, bus, subway or other means of transportation).

4.9 Big data for smart transportation [18]

Data are generated from many sources in the form of big data. Data sources are available devices like environmental sensors, computers, smartphones, cameras, GPS and so forth. Various applications, for example digital pictures and videos, advertising applications, social media sites and commercial transactions, have accelerated data generation in the past few years [22, 23].

Figure 4.3 depicts that connected city can be classified in different aspects and fields like healthcare, transportation, power grid, smart education, etc. These systems are connected with big data system and transfer data to each other.

Implementing big data in smart cities is very challenging in dynamic environments [17]. Some of the challenges are described in what follows.

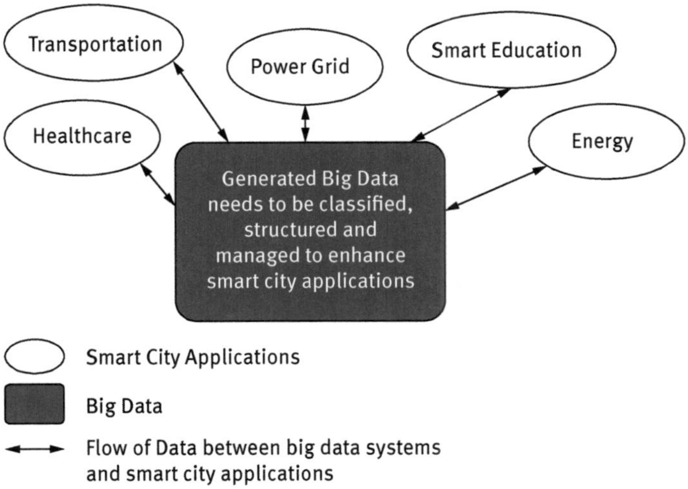

Fig. 4.3: Classification of connected cities

4.9.1 Data size and sources

In smart cities, data are generated from various sources in various formats. Some of these are structured, some may be unstructured or semi-structured. Data storage and data processing are not easy tasks using traditional software.

4.9.2 Data information and sharing

Smart city applications deal with data integration from multiple organizations, heterogeneous environments and a variety of intelligent devices. Data integration of an organization is one of the biggest challenges in the IT field. Integration of data from many departments becomes challenging. There are also challenges in sharing information with many departments. Each department maintains its own data warehouse, so sharing information is difficult.

4.9.3 Data quality

Data in connected cities are collected from different heterogeneous systems and in various formats, so maintaining the quality of data becomes challenging. Different transactional systems need data in a similar format. But when data arrive at a big data warehouse, where the data from various sources are gathered, they need to be in specific formats, without consideration for their quality. Thus, it is necessary to ensure that proper mechanisms are in place to retrieve quality data.

4.9.4 Data security

Data security is an important issue in dealing with big data. Big data faces more serious damage than normal size data. Data movement from various sources to various destinations in a secure manner is one of the great challenges for connected city applications. Security means information should go to the right person and the right place. To design a connected city, data masking and security are also important issues so that data do not end up in the wrong hands. Nowadays, advanced technologies are very complex and costly. The connected city should be designed such that a better solution is arrived at than is possible with existing solutions.

4.9.5 Lack of skills

A lack of skills in dealing with the variety of data sometimes becomes an obstacle to effectively using big data in connected cities. Having the skills needed to manage and analyse large amounts of data is very important. Developing an effective policy to improve such skills is necessary.

4.9.6 Lack of standards

A lack of standards is another aspect to consider. For the development of a connected city, some standards should be maintained that will play a significant role in settling on a common platform. A lack of standards affects data formats, data integration and procedures for data mining, as well as other policies related to data.

4.10 Big data applications in the connected city

Application of big data technologies enables data storage to process information that serves different connected cities. Big data also helps to make decisions on any expansion of services and resources in a smart city. To achieve the goals and services for connected cities, big data requires tools and methodologies for effective data analysis. Tools and methodologies collaborate and communicate between different entities and provide various services for connected cities. A three-layer management system is proposed [24] to support smart urban mobility with an emphasis on bus transportation. The development of location tracking technology in transportation systems significantly increases the amount and variety of data. Predictive modelling of bus arrival times [25, 26] improves passenger experience. An intelligent transportation management system, by investing in AVLAPC-(Automatic Vehicle Location (AVL) Automatic Passenger Counters (APCs))systems [27], has also been proposed in research work. Predictions are made based on a model of time series like autoregressive integrated moving average (ARIMA), seasonal ARIMA (SARIMA) or a hybrid approach [28, 29]. These types of predictions are made to forecast passenger demands. Passenger demands are modelled in an aggregated way, so individual transit patterns are not considered. Another limitation is that these approaches are built on a limited number of routes. As a result, passenger transfer behaviour is usually ignored. Other important prediction methodologies have also been established to predict bus arrival times [25, 26]. Nowadays, the research work on smart transportation systems includes, for example, smartphone applications to collect various kinds of information like bus trajectories and passengers' on-board experience via crowd sensing [30]. This approach does not depend on special bus equipment. However its data quality largely depends on how many users are motivated to download and use the app. Also, the collected data may be biased due to self-reporting errors.

4.10.1 Smart grid

The rapid distribution of smart grids has enabled researchers to analyse and use real-time consumption of generated data. Improvements in intelligent services directly enhance the results in existing smart grid infrastructures. In a smart grid environment, huge amounts of data are generated from various sources for energy consumption. Big data collected from a smart grid environment helps in making decisions about the supply level of electricity to meet user demands. Smart grid data analytics helps to predict future power supply needs.

4.10.2 Smart health care

The world's population is increasing rapidly. Collection and analysis of patients' data by specialists require efficient tools. Such tools are also used by administrative organizations and insurance agencies. Big health care data should be properly analysed to predict diseases and their cures. They also improve the quality of life and reduce death rates. Data or information collection for a particular patient is done by intelligent gadgets. The gadgets are associated with clinics or hospitals to help keep patient records.

4.10.3 Smart transportation

The patterns of large amounts of traffic data improve transportation systems to minimize traffic congestion by advertising alternative routes and reducing the number of accidents. Big data gathered from smart transport systems can help shipments and shipping movements by reducing supply chain wastage. Data of smart transport also provide several benefits like reducing environmental impact and increasing safety.

4.10.4 Smart city governance

Big data analytics plays an important role in smart governance. Different organizations with common goals can easily identify data to be analysed using data analytics with collaborating organizations. Different countries collaborate with each other for their development. Big data analytics helps governments establish and implement policies in such areas as education, social care and health, for example. The rate of implementation of big data can be increased by analyzing large amounts of data from several institutes.

4.11 Big data and data analytics [31]

Connected cities generate huge amounts of data just over the course of their daily operations. To gather additional amounts of data, the IoT and Open Data are used for smart cities. Some data are static, some are real-time. These data show the classic characteristics of big data: high volume, real time and heterogeneous in format, sources and characteristics. Big data can be managed and analysed well and can offer economic value that cities can use to enhance efficiency. Big data can also lead to new ideas that enhance the quality of urban dwellers. Cities can access and use many computing resources that are very expensive to manage. Technologies like HDFS, Hive, Hadoop and Spark are now available for harnessing big data and other analytical tools to enhance the efficiency of city operations and services.

Big data has made drastic changes in various fields including ITS. The emerging technologies created around ubiquitous digital devices have opened up several opportunities for enhancing the performance of ITSs. However, the magnitude and heterogeneity of big data are beyond the capabilities of existing approaches to ITSs. The proposed architecture [32] is based on systematic analysis of the requirements of existing traffic control systems. The big data analytics engine informs the control logic. While making a system for data analytics, the variability of queries should be taken into account. According to [32], the possible queries are divided into four groups: periodic vs. non-periodic, descriptive vs. predictive and real-time vs. non-real-time, single vs. multiple sources. The paper also proposed an architecture depending upon some set of requirements of a given traffic control system.

Advantages of big data for the transportation industry are as follows:
- A traffic management system used for big data technology deals with large volumes of data. A large job is divided into small jobs and completed in a MapReduce model. The stability and fault tolerance of the model is also important.
- Big data can improve the performance level of transportation. The transportation industry, which involves many fields of work that are necessary to handle huge volumes of data, also uses other models to control applications. Big data has good forecasting capabilities. Big data reduces the chances of false alarms and underreporting of various incidents related to traffic. Traffic control is a vital part of ITSs. Regarding travel-related information, an ITS can inform travellers about traffic conditions and empower them to make decisions concerning the best route to improve the traffic situation in a city. To improve the efficiency and capacity of a transportation road network, big data technology plays a major role.
- Big data also improves traffic safety. Big data can be used by real-time processing systems to identify traffic accidents and effectively predict them using video surveillance, microwave detection and mobile detection systems. When security incidents happen and emergency conditions arise, big data can improve conditions for emergency responses.

Intelligent transportation on a big data platform represents a collection of multiple technologies, models, systems and departments. The system consists of management, science, mathematics, economics and information technology. The system also includes a data analysis layer, a basic business layer and an information-publishing layer.

The basic business layer is a platform for a data analysis layer and then an information-publishing layer. The main activity of this layer is to finish the basic job of the different business units and produce data in the basic business layer. The layer contains control systems, a traffic information collection system, a video surveillance system, signals, a forensic crime system and others. The basic business layer's task is to deal with the different business organizations. In the basic business layer, cloud computing technology is used. The security and stability of the application system is ensured by the integration of a decentralized system with the cloud, which provides an efficient computing environment. The data analysis layer of the given architecture uses data mining technology, big data technology, which is combined with various mathematical models for real-time analysis. The layer can be used for traffic guidance, congestion warnings and other intelligent transportation purposes. The data analysis layer is created in a Hadoop ecosystem. It uses a cheap commercial server, the open-source Linux system for big data storage. The information-publishing layer provides traffic conditions to, for example, travellers, industry executives and other business units via desktop applications, mobile terminals, the Internet and other applications.

4.12 Conclusion

This chapter surveyed the smart transportation of a connected city. The usefulness of big data for a connected city is discussed. Various applications related to big data, such as IoT, are discussed. There is also a discussion on how big data and IoT are involved in smart transportation, especially with respect to traffic congestion management, travel help and other scenarios. Big data and the IoT are very useful technologies nowadays. These two technologies are having a significant impact on smart transportation in smart or connected cities. The chapter describes the relationship between big data and IoT and smart transportation in connected cities. In the future, many emerging technologies will also improve smart transportation service and efficiency. The challenges of big data are another important issue to address in future research.

Bibliography

[1] Wireless Communications in Transportation Systems. *Hindawi.*

[2] Amini S, Gerostathopoulos I, and Prehofer C. Big Data Analytics Architecture for Real-Time Traffic Control. *TUM Living Lab Connected Mobility (TUM LLCM) project Center Digitisation.* Bavaria, Bavarian State Government.

[3] Zheng X, Chen W, Wang P, Shen D, Chen S, Wang X, and Zhang Q. Big Data for Social Transportation. *IEEE Transactions on Intelligent Transportation Systems*, 17(3), 2016.

[4] Wang X and Li Z. Traffic and Transportation Smart with Cloud Computing on Big Data. *International Journal of Computer Science and Applications*, 13(1):1–16, 2016.

[5] Al Nuaimi E, Al Neyadi H, and Al-Jaroodi NMJ. Applications of big data to smart cities. *Journal of Internet Services and Applications*, 2015.

[6] Fan W and Bifet A. *Mining big data: current status and forecast to the future.* ACM SIGKDD ExplorNewsl., 2013.

[7] Michalik P, Stofa J, and Zolotova I. Concept definition for Big Data architecture in the education system. In *Applied Machine Intelligence and Informatics (SAMI), IEEE 12th International Symposium*, pages 331–334, 2014.

[8] Khan M, Uddin MF, and Gupta N. Seven V's of Big Data understanding Big Data to extract value. *IEEE*, pages 1–5, 2014.

[9] Khan Z, Anjum A, and Kiani SL. Cloud Based Big Data Analytics for Smart Future Cities. In *Proceedings of the 2013 IEEE/ACM 6th International Conference on Utility and Cloud Computing*, pages 381–386, 2013.

[10] Su K, Li J, and Fu H. Smart city and the applications. *Electronics, Communications and Control (ICECC), IEEE*, pages 1028–1031, 2011.

[11] Lee CH, Birch D, Wu C, Silva D, Tsinalis O, Li Y, and Guo Y. Building a generic platform for big sensor data application. *IEEE International Conference*, pages 94–102, 2013.

[12] Kim GH, Trimi S, and Chung JH. Big-data applications in the government sector. *Commun, ACM.*, 57(3):78–85, 2014.

[13] Kramers A, Höjer M, Lövehagen N, and Wangel J. Smart sustainable cities – Exploring ICT solutions for reduced energy use in cities. *Environ Model Software*, 56:52–62, 2014.

[14] Aguilera G, Galan JL, Campos JC, and Rodríguez P. *An Accelerated-Time Simulation for Traffic Flow in a Smart City*. FEMTEC, 2013.
[15] Neirotti P, De Marco A, Cagliano AC, Mangano G, and Scorrano F. *Current trends in Smart City initiatives: Some stylised facts*. Cities, 2014.
[16] Mohamed N and Al-Jaroodi J. Real-time big data analytics: Applications and challenges. In *International Conference on High Performance Computing and Simulation (HPCS)*, pages 305–310, 2014.
[17] Kumar S and Prakash A. Role of Big Data and Analytics in Smart Cities. *Journal of Science and Research (IJSR)*, 5(2), February 2016.
[18] The Smart/Connected City and Its Implications for Connected Transportation, 2014. *White Paper*. October 14.
[19] Lohr S. The Internet Gets Physical. *The New York Times*, December 17, 2011.
[20] Wasik B. In the Programmable World. All Our Objects Will Act as One. *Wired*, May 14, 2013.
[21] Morioka M, Kuramochi K, Mishina Y, Akiyama T, and Taniguchi N. City Management Platform Using Big Data from People and Traffic Flows. *Hitachi Review*, 64(1), 2015.
[22] Harrison C and Donnelly A. A Theory of Smart Cities, 2011. *White Paper*.
[23] Smart Cities Council.
[24] Wang Y, Ram S, Currim F, Dantas E, and Sabóia LA. *A Big Data Approach for Smart Transportation Management on Bus Network*. IEEE, 2016.
[25] Bai C, Peng Z, Lu Q, and Sun J. Dynamic Bus Travel Time Prediction Models on Road with Multiple Bus Routes. *Comput. Intell. Neurosci*, pages 1–9, 2015.
[26] Tétreault PR and El-Geneidy AM. Estimating bus run times for new limited-stop service using archived AVL and APC data. *Transp. Res. Part A Policy Pract.*, 44(6):390–402, 2010.
[27] Furth PG, Hemily B, Muller THJ, and Strathman JG. Uses of archived AVL-APC data to improve transit performance and management: Review and potential. *Transp. Res. Board*, 23, no. June, 2003.
[28] Xue R, Sun DJ, and Chen S. Short-term bus passenger demand prediction based on time series model and interactive multiple model approach. *Discret. Dyn. Nat. Soc*, i:1–11, 2015.
[29] Moreira-Matias L, Gama J, Ferreira M, Mendes-Moreira J, and Damas L. Predicting taxi-passenger demand using streaming data. *IEEE Trans. Intell. Transp. Syst*, 14(3):1393–1402, 2013.
[30] Farkas K, Feher G, Benczur A, and Sidlo C. Crowdsending based public transport information service in smart cities. *IEEE Commun. Mag*, 53(8):158–165, 2015.
[31] Lea R. Smart Cities: An Overview of the Technology. *Trends Driving Smart Cities IEEE*, March, 2017.
[32] Amini S, Gerostathopoulos I, and Prehofer C. Big data analytics architecture for real-time traffic control. *Models and Technologies for Intelligent Transportation Systems (MT-ITS), IEEE*, June, 2017.

S. V. Gorbachev

5 Model of intellectual analysis of multidimensional semi-structured data based on deep neuro-fuzzy networks

Abstract: A new structure of deep neuro-fuzzy networks is proposed and studied, combining through a layer of fuzzy clustering a Kohonen fuzzy cellular neural network self-organizing map (FCNN-SOM) and a radial basis function network (RBFN). The proposed model has a high degree of self-organization of neurons, improving the separation properties of the network in the case of overlapping clusters; automatic adjustment of the parameters of radially symmetric functions; a single hidden layer sufficient for modelling pronounced non-linear dependencies; simplicity of the algorithm for optimizing weight coefficients; and a fast learning speed. It can be used to solve a wide range of problems – clusterization, approximation and classification (recognition) of multidimensional semi-structured data.

Keywords: Semi-structured data, Analysis, Kohonen fuzzy cellular neural network, Radial basis function network

Introduction

In recent years, the field of big data has emerged as one of the most promising in computer technology. Taking into account the specifics of the growth in technical progress, it is easily explained – digital technologies accompany many aspects of modern life and constantly gather information. The amount of data on various aspects of life is increasing, along with growing possibilities of information storage. According to a study by IDC Digital Universe, in 2020 the amount of data in the world can reach 40 ZB (zettabytes), which is the equivalent of 5,200 GB for every person on the planet (Figure 5.1).

At this rate of growth, the amount of data in the world, according to forecasts by researchers, will double annually. The big data paradigm identifies three main types of tasks:

1. Storing and managing hundreds of terabytes or petabytes of data; a relational database is not able to work effectively with this amount of information;
2. Organizing unstructured information consisting of texts, images, video and other data types;

S. V. Gorbachev, National research Tomsk State University, Tomsk, Russia, email:spp03@sibmail.com

https://doi.org/10.1515/9783110551433-005

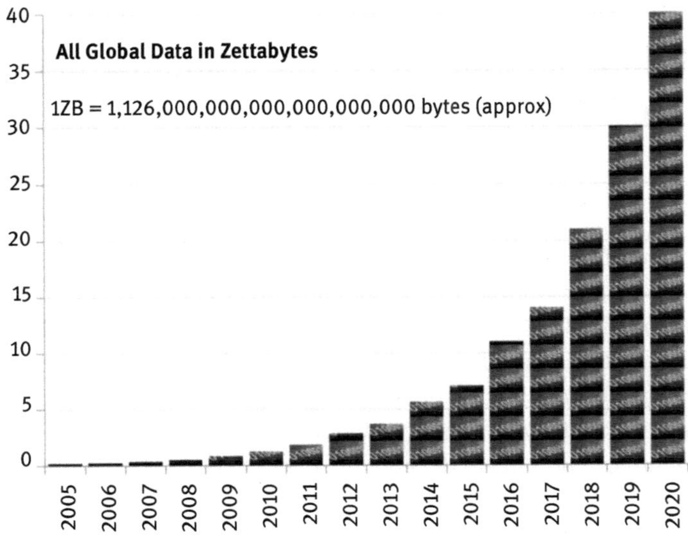

Fig. 5.1: Statistical data volumes are increasing, with a forecast up to 2020

3. Analysing big data, which raises the question of how to work with unstructured and substructural information, including clustering methods, approximation and classification (recognition) of multidimensional semi-structured data, the generation of analytical reports and implementation of predictive models.

Despite the apparent simplicity of the approaches to big data, their application has several features. One of the main factors that hinders the implementation of big data projects, in addition to the high cost, is the problem of selecting the data to be processed, i.e., determining which data to retrieve, store and analyse and which to leave out. Another problem of data processing is that the high computational complexity of algorithms is associated with a large number of parameters and attributes.

This chapter examines one of the possible ideas about applying the paradigm of big data – the ability to create high-performance intellectual models for multidimensional semi-structured data based on soft computing algorithms.

This work's purpose is to develop and research a high-performance and at the same time compact deep neuro-fuzzy classifier that is able to perform a wide range of tasks in the fuzzy clustering, approximation and classification (recognition) of multidimensional semi-structured data.

To achieve this goal it is necessary to solve the following problems:

1. Analysis of methods and approaches to solving problems of clustering, approximation and classification (recognition), identifying their advantages and disadvantages;
2. Development of a conceptual model of analysis;

3. Research of neural network classifier based on a radial basis function network (RBFN);
4. Development architecture and learning algorithms of Kohonen fuzzy cellular neural network self-organizing map (FCNN-SOM).
5. Development of a model of a generalized deep fuzzy-neural classifier;
6. Effective justification of results – qualitative and quantitative assessment of the effectiveness of the model and approach in general, analysis comparing the model with classical neural network models.

This chapter contains seven sections. The first section provides an overview of the subject area, an analysis of the publications on the research topic, a comparison of the existing approaches to the analysis of data, identifying their advantages and disadvantages. The second section provides a justification of the chosen conceptual model of analysis and the corresponding neural net and fuzzy methods. The third, fourth and fifth sections discuss the implementation details of a new deep neuro-fuzzy model based on the proposed theoretical approach and provide a description and mathematical foundation for the algorithms and methods. The sixth section is devoted to the experimental verification of the obtained results on the test data. The seventh section describes the practical results of applying the developed model to build the trajectory of world economic and technological development and a calculation of forecast parameters of the level and pace of economic and technological development in several countries. The conclusion summarizes the results of the work performed.

5.1 Literature review

Recognition is the problem of constructing and applying formal operations on numeric or symbolic representations of objects, in a real or ideal world; the outcomes of decisions of the objects reflect the relationship of equivalence between them. The relations of equivalence express the membership of the evaluated objects in any of the classes, considered as independent semantic units.

The classes of equivalence can be defined by a researcher, while constructing recognition algorithms, who uses his own substantive views or uses external information about the similarities and differences between objects in the context of the problem being solved. In this case, the term “pattern recognition with a teacher” is used [1]. Otherwise, when an automated system solves the classification problem without engaging external teaching information, it is referred to as automatic classification or “pattern recognition without a teacher”.

Various authors (e.g., Y. L. Barabash, V. I. Vasiliev, A. L. Gorelik, V. A. Skripkin, R. Duda, P. Hart, L. T. Kuzin, F. I. Peregudov, F. P. Tarasenko, F. E. Temnikov, John. Tu, R. Gonzalez, P. Winston, K. Fu, Y. Z. Tsypkin) have proposed different classification methods of pattern recognition. Some authors distinguish between parametric, non-parametric and heuristic methods, while others isolate groups of methods based on historical schools and streams in this area. For example, in the work of V. A. Duke [2], in which an academic overview of recognition methods is given, the following typology is used:

- Methods based on the principle of separation;
- Statistical methods;
- Methods based on "potential functions";
- Methods of computing ratings (voting);
- Methods based on propositional calculus, in particular on the unit of the algebra of logic.

This classification is based on the difference in formal methods of pattern recognition and therefore the heuristic approach to recognition, which has received full and adequate development in expert systems, is not considered here. The heuristic approach is based on knowledge that is difficult to formalize and on the researcher's intuition. The researcher decides which information will be used and how the system should use it to achieve the desired effect of recognition.

Such a typology of recognition methods with varying degrees of detail is found in many works on recognition. At the same time, well-known typologies do not take into account one very significant characteristic that reflects the specificity of the method for representing domain knowledge with the help of any formal algorithm of image recognition.

D. A. Pospelov in [3] distinguishes two main ways of representing knowledge:

- Intensional, in the form of diagrams of relationships between attributes (characteristics);
- Extensional, with specific facts (objects, examples).

Intensional representation captures patterns and connections that explain the structure of data. In the context of diagnostic tasks, such capture is to define operations on the attributes (characteristics) of objects, leading to the desired diagnostic result. Intensional representations are implemented by operations on attribute values and are not intended to operate on particular information with facts (objects).

Conversely, extensional representation of knowledge is associated with a description and fixation of particular objects of the subject area and implemented in operations whose elements are objects as a whole system.

One can draw an analogy between intensional and extensional knowledge and the mechanisms underlying the activities of the left and right hemispheres of the human brain. If the right hemisphere is characterized by the prototype of a holistic representation of the world, the left hemisphere deals with patterns that reflect the relationship between the attributes of this world.

Based on the two foregoing fundamental ways of representing knowledge, Lutsenko has suggested the following classification of methods of pattern recognition [4, 5]:
- Intensional methods based on operation characteristics;
- Extensional techniques based on operations with objects.

It should be stressed that the existence of these two (and only two) groups of methods of recognition (operating with characteristics and operating with objects) is deeply regular. From this point of view, neither of these methods, taken separately from the other, provides an adequate reflection of the subject area. According to the author, between these methods there is a relation of complementarity in the sense of Niels Bohr [6], so a promising recognition system needs to provide an implementation of both methods, but not just one of them.

Thus, it is possible to draw the following conclusions:
1. The basis of classification of pattern-recognition methods is built on fundamental laws underlying the human method of cognition in general, which puts it in a very special (privileged) position compared to other classification methods, which seem more lightweight and artificial.
2. An overview of recognition methods reveals the following contradiction. Currently, a number of different methods of pattern recognition and classification is theoretically developed and described in the literature. However, a software implementation of most methods is lacking and it is highly regular, it might even be said to be predefined by characteristics of the recognition methods themselves. This is evidenced by the fact that such systems receive little coverage in the literature and other sources of information.
3. Consequently, the question of the practical applicability of certain theoretical methods of pattern recognition, which are designed to solve practical problems in real (i.e., quite large) dimensions of data and on real, modern computers remains insufficiently explored.

The aforementioned facts can be understood if it is recalled that the complexity of mathematical models exponentially increases the complexity of software implementation of a system and, to the same extent, reduces the chances that the system will work in practice. This means that one can really implement on the market only those software systems that are based on rather simple and transparent mathematical models. Therefore, developers interested in the replication of their software should consider the choice of mathematical model not only from a scientific point of view but from a pragmatic one as well, given the capabilities of the software being implemented. Models should be as simple as possible and, thus, be developed at less cost and more efficiently and work (be practically effective).

In this regard, the task of actuating the mechanism of generalizing descriptions of objects belonging to a certain class in recognition systems is particularly relevant, i.e., the mechanism of forming compact generalized images. Obviously, such a generalization mechanism will make it possible to "compress" any dimension of the training sample to a previously known of dimension generalized image basis. It will also pose and solve a number of tasks that cannot even be formulated in the following recognition methods – the method of comparison with the prototype, the method of k-nearest neighbours and the algorithm of calculating estimates.

That are the tasks of:

- Determine the information contribution of the signs in the information portrait of a generalized image;
- Perform a cluster-constructive analysis of generalized images;
- Determine the semantic load of attributes;
- Perform a cluster semantic-structural analysis of signs;
- Make a meaningful comparison of the generalized images of classes with each other and signs with each other (cognitive chart, including the chart Merlin).

In practice, original data are often difficult to formalize and often have an inhomogeneous structure with deliberately overlapping classes of images. Another problem is the incompleteness of training samples and their strong noise. In a discriminant analysis in cases of uncertainty about responses on images belonging to a certain class of images, the answer can be obtained in the form of a probability of belonging to each of the image classes. However, the previously described features of data in many cases do not allow one to construct adequate probabilistic and statistical models that motivate the creation of empirical methods and approaches, and during their development methods and models from the field of soft computing become convenient.

Methods of fuzzy logic and neural network technology currently relate to the perspective of adaptive technologies of information processing and solving problems of pattern recognition and prediction, allowing one to create high-quality intelligent systems in conditions of multidimensional objects of intersecting classes analysis [7–9].

A significant scientific contribution to the theory and practice of creating diagnostic test systems built on fuzzy logic and neural networks was introduced by the following scientists: L. Zadeh, D. A. Pospelov, Y. A. Bortsov, F. Wasserman, N. Hassoun, A. P. Rotshtein, V. I. Gostev, A. I. Galushkin, V. V. Kruglov, D. Rutkowski, M. Pilinsky, S. Omatu and others [10–15].

An artificial neural network based on radial-symmetric (radial basis) functions (Figure 5.2), or an RBFN network, can be used to solve a wide range of tasks, among which the most commonly used are approximation, classification and data clustering [9, 16, 17].

The known methods for constructing RBFNs [12, 18] can be divided into two groups. The methods of the first group assume that the number of RBFN neurons in the first layer is set by the user, and the number of neurons in the second layer is determined by the network output dimension. After that, the RBFN is treated as a special case of a multilayer neural network of a direct signal propagation (multilayer perceptron), which is taught using gradient methods of multidimensional non-linear unconditional optimization [18], and the partial derivatives of the objective function are determined on the basis of a reverse error propagation technique [18]. In this case, there is uncertainty in the choice of the number of neurons in the first layer, which can lead to an overflow network or, conversely, to the impossibility of constructing a model with the required accuracy. Another disadvantage of these methods is the uncertainty in the choice of the initial values of the network weights, which can lead to an inability to solve the task of training the RBFN for a limited time.

The methods of the second group assume that the training sample is displayed in the RBFN structure. This, as a rule, is done by simple memorization of the entire sample or based on cluster analysis, which determines the number and coordinates of the cluster centres that are stored in the RBFN memory. The number of neurons in the first layer of the RBFN is set equal to the number of clusters, and the coordinates of the cluster centres are entered in the weights of the first layer's neurons. Further, the RBFN is trained to correct the weights of neurons of the second layer on the basis of the methods of multidimensional non-linear unconditional optimization [12, 18]. If the network is obtained by displaying the sample, then, as a rule, it does not provide a generalization. If a cluster analysis is used, the quality of the obtained neuro-model essentially depends on the quality of the results of the cluster analysis; the network can be superfluous and exhibit low generalizing properties due to excessive granularity of the partition of the feature space. To solve this urgent problem, we developed a special architecture for the Kohonen FCNN.

The traditional approach to the performance of these tasks assumes that each observation can belong to only one cluster. In this case, the most widely used are Kohonen neural networks [1], which have a single-layer architecture with lateral connections and which are trained on the basis of so-called the winner takes all (WTA) or winner takes most (WTM) rule. The more natural situation is one where the processed feature vector with different levels of conditioning (probability, possibility) can belong to multiple classes. This situation is the subject of fuzzy cluster analysis, which developed a self-learning deep neuro-fuzzy model, which is a generalization of the Kohonen neural network and has, due to the use of special algorithms for the adjustment of its semantic weights, more functional opportunities. Thus, a modification of Kohonen network was introduced in [19, 20] based on fuzzy rules. This network has shown its effectiveness in a number of tasks associated with recognition; however, the associated numerical complexity hinders its practical use. In [21, 22] was proposed a fuzzy Kohonen network for clustering, which essentially uses a C-means clustering method. In [10], a Kohonen network with fuzzy inference trained on the basis of the combination of Kohonen and Grossberg rules was proposed. The main disadvantage of this design is the dependence of the results on the choice of the free parameters of the training procedure. In [23] the possibility of training a Kohonen artificial neural network by a cellular automaton (CNN-SOM) was proved, which can greatly enhance the quality and speed of training; however, the issues related to separating the properties of CNN-SOMs in the case of overlapping clusters have not been resolved.

5.2 A conceptual model of analysis

In this chapter we introduce a new two-layer Kohonen (FCNN-SOM) that is an adaptive modification of the FKSN [19] and the further development of the composition in [23] and the algorithms for its learning.

Next, on the basis of combining FCNN-SOM and a RBFN, we describe a new deep neuro-fuzzy model for clustering and classification (pattern recognition). The concept of this model is based on the classification by Pospelov [1]; the two ways of representing knowledge are extensional (the formation of clusters of generalized patterns) and intensional (identifying integral characteristics in the selected clusters of classes).

Note that most pattern recognition algorithms require the involvement of large computing power, which can only be achieved on high-performance computer equipment. Therefore, an important task is to build a compact network architecture.

5.3 Research on neural network classifier based on RBFN

An artificial neural network based on radial-symmetric (radial basis) functions (Figure 5.2), or RBFN, can be used to perform a wide range of tasks, the most frequent of which are approximation, classification and data clustering.

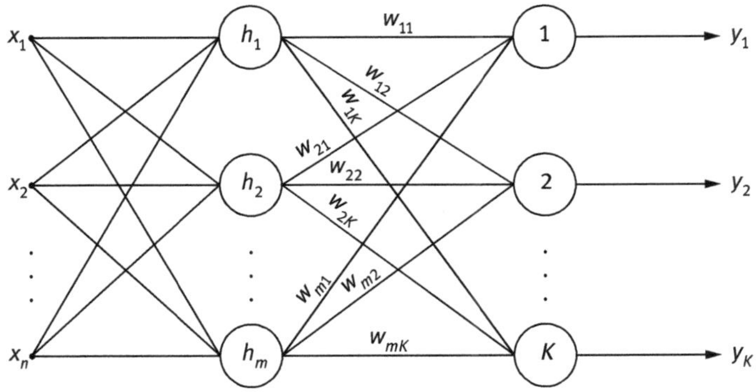

Fig. 5.2: Architecture of neural network based on radial basis functions

The main property of radially symmetric functions that they are monotone and symmetric about some vertical axis of symmetry and their changing (decrease or increase) responses. An example of such a function is the expression of the Gaussian function

$$f(s) = \exp(-\alpha(s - T)^2). \tag{5.1}$$

This function is most often used in neural network architectures, but mainly in a multidimensional case:

$$\vec{h}(x) = \exp(-\alpha\|x - \vec{c}\|^2)\,, \tag{5.2}$$

where $\vec{c}$ is the vector of the centres (the coordinates of the vertical axes of symmetry) of the set of radially symmetric functions;

$\|x - \vec{c}\|$ is the norm of the deviation vector of the input variable from the centres of radially symmetric functions. The parameter α is related to the scattering radius r of the input variables and can be replaced by the corresponding ratio:

$$\alpha = \frac{1}{2r^2}\,. \tag{5.3}$$

The norm of the vector difference is calculated as the Euclidean distance:

$$\|x - \vec{c}\| = \sqrt{(x - c_1)^2 + \ldots + (x - c_m)^2}\,. \tag{5.4}$$

The RBFN architecture contains three layers of neurons (Figure 5.2). The first layer is the input layer. Outputs of the second (hidden) layer are activated by a set of radially symmetric functions. In fact, they process the vector of input values, determining the degree of proximity of each of them to the centres of radially symmetric functions. Outputs of neurons of the third layer (i.e., outputs of the entire neural network) are linear combinations of the outputs of the second layer.

The composition and number of inputs and outputs are determined by the class of the problem being solved. When data are approximated, the inputs are the arguments of the approximating relationship, and the outputs are the values returned by it. When clustering or classifying data, inputs are characteristics that distinguish objects classified as clusters or classes, and outputs indicate a cluster or class corresponding to inputs.

The number of hidden elements also depends on the task being solved. If this is an approximation of the data, it can be any number. In the case of clustering or classifying data, it should correspond to the number of clusters or reference images of classes.

The life cycle of artificial neural networks based on radially symmetric functions, as for most other architectures, includes two stages: training and practical use. In turn, at the learning stage, two stages can also be distinguished: tuning the neural network and optimizing the synaptic coefficients of the linear output layer.

At the stage of tuning the neural network under consideration, it is necessary to determine the centres c and radius r of the radial elements (neurons of the hidden layer). The following settings are possible.

1. In the presence of a small number of reference samples for training as centres of radially symmetric functions, the vectors corresponding to them are usually chosen. If the volume of the training sample is large enough, the following can be used as centres:
   - The centres of potential clusters, by which you can distribute all examples of the training sample manually or using additional clustering algorithms, including other neural network architectures;
   - Individual random examples of the training sample.

   It should be noted that the second option is better to use with a large number of neurons in the hidden layer.
2. The choice of the radius of the radial elements is determined by the required form of the radially symmetric function. For large values of the parameter α, the graph of the function is too sharp, which means that the network will not correctly interpolate the data between known points at a sufficiently large distance from them, as it loses the ability to generalize the training data. Conversely, if the parameter α is too small, the network becomes unresponsive to the individual parts.

   In view of the foregoing, the radius can be specified in the following ways:
   - By a user of the neural network in an explicit form based on the heuristic selection;
   - By automatically calculating it as the average distance to a few (depending on the total training sample size and the number of hidden neurons) of the nearest examples.

RBFN networks have the property of fast convergence and can be trained more quickly on the basis of many examples in comparison with neural networks trained by the backpropagation algorithm of the error.

At the step of optimizing the weight coefficients of the linear output layer, the following steps are sequentially performed:

1. The characteristic matrix of the values of the radially symmetric elements of all training samples is calculated:

$$\overline{\overline{H}} = \begin{bmatrix} h_1(x^{(1)}) & h_2(x^{(1)}) & \dots & h_m(x^{(1)}) \\ h_1(x^{(2)}) & h_2(x^{(2)}) & \dots & h_m(x^{(2)}) \\ \dots & \dots & \dots & \dots \\ h_1(x^{(N)}) & h_2(x^{(N)}) & \dots & h_m(x^{(N)}) \end{bmatrix} \tag{5.5}$$

   The number of rows of this matrix is equal to the number of examples of learning sample. The number of columns is the number of radial elements.

2. The methods of linear algebra are use to calculate the matrix of weight coefficients of the output layer neurons:

$$\overline{\overline{W}} = (\overline{\overline{H}}^T \cdot \overline{\overline{H}})^{-1} \cdot \overline{\overline{H}}^T \cdot \overline{\overline{Y}}\,, \tag{5.6}$$

where the output matrix of training samples contains columns in an amount equal to the number of training samples, and the number of strings corresponding to the number of outputs of the neural network:

$$\overline{\overline{Y}} = \begin{bmatrix} y_{11} & y_{21} & \cdots & y_{k1} \\ y_{12} & y_{22} & \cdots & y_{k2} \\ \cdots & \cdots & \cdots & \cdots \\ y_{1N} & y_{2N} & \cdots & y_{kN} \end{bmatrix} \tag{5.7}$$

Among the advantages of the considered architecture of neural networks are the following:
- The presence of a single hidden layer, sufficient for modelling pronounced nonlinear dependencies;
- Simplicity of the algorithm for optimizing weight coefficients;
- Guaranteed to find the global optimum of the error function when finding the weight coefficients of the neurons of the output layer;
- Fast learning speed.

The limitations or disadvantages of neural networks based on radially symmetric functions include the following:
- The need for special adjustment of the parameters of radially symmetric functions and the complexity of tuning with a large number of hidden radial elements;
- The impossibility of extrapolating the model beyond the initial interval of changing the input values of the training sample.

If the network is obtained by displaying a sample, then, as a rule, it does not provide a generalization. If a cluster analysis is used, the quality of the obtained neuro model essentially depends on the quality of the results of the cluster analysis, the network can be superfluous and exhibit low generalizing properties due to the excessive granularity of the partition of the feature space. To solve this urgent problem, we developed a special architecture for the Kohonen FCNN. Its output layer of fuzzy clustering simulates the symmetrical bell-shaped functions of the input example belonging to each of the generalized clusters. In the next section, we show the proposed architecture and describe algorithms for calculating and adjusting the parameters of the centres and the widths of these functions. The final result of this is our proposal to combine the output layer of the FCNN with the second (hidden) layer of the RBFN and thereby solve the most difficult stage of the RBFN setup – partitioning the input attribute spaces, selecting a view and setting the parameters of radially symmetric functions.

5.4 Development of architecture and learning algorithms for Kohonen FCNN

Phase one: Self-organization

Within the extensional approach, we propose to perform a segmentation into clusters of certain classes of patterns by Kohonen SOMs [1] so as to solve the problem of separating properties in the case of overlapping clusters. This will make it possible to improve recognition quality.

A SOM has a simple architecture, including zero receptor layer and only one layer of Kohonen neurons, which are adaptive linear adders such as a rectangular grid in a plane (Figure 5.3). Each j-th neuron ($j = 1, \ldots, m$) is characterized by an n-dimensional vector of synaptic weights $W_j = (w_{j1}, w_{j2}, \ldots, w_{jn})^T$.

The input vectors $x(k) = (x_1(k), \ldots, x_n(k))^T$, where k is the number of the example in the training set, are sequentially distributed from the receptive (zero) layer to all the neurons N_j of the Kohonen layer; in addition, their synaptic weights $w_{ji}(k)$ define the centroids m of overlapping clusters.

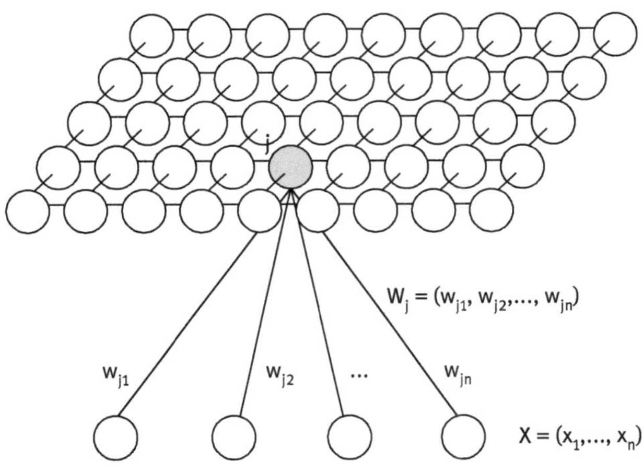

Fig. 5.3: Model of Kohonen self-organizing networks

The basis for the self-organization of neural networks is the observed regularity that the global ordering of networks becomes possible as a result of self-organizing operations, conducted independently of each other in various local segments of the networks. In accordance with the submitted signals, neurons are activated. Finally, one neuron becomes active in a network (or in a group). The output neuron that won the competition is called the winner neuron.

Neurons in the course of the competitive process, due to changes in the values of the synaptic weights, are selectively tuned to different input vectors or classes of input vectors. In the learning process there is a tendency to increase the values of the weights, because of which a peculiar positive feedback is created: more powerful exciting impulses – higher values of weights – greater activity of neurons.

In this case, there is a natural stratification of neurons into different groups; individual neurons or their groups cooperate with each other and they are activated in response to excitation created by specific training vectors, suppressing other neurons by their activity. One can speak about cooperation between neurons within a group and about competition between neurons within a group and between different groups.

Among the mechanisms of self-organization, there are two main classes: self-organization based on the associative Hebb's rule, and the mechanism of neuron competition based on the generalized Kohonen rule. Henceforth, we will consider the mechanism of neuron competition.

The formation of self-organizing networks begins with the initialization of the synaptic weights of the network. Usually, synaptic weights are assigned small values that are generated by a random number generator. During this initialization, the network initially will have no order of signs of the input vectors. After the initialization of the network, three basic processes are implemented [1]:

1. Competition: For each input vector, network neurons calculate the relative values of the discriminant function;
2. Cooperation: The winning neuron determines the topological neighbourhood of a group of neurons, providing a basis for cooperation between them;
3. Synaptic adaptation: Correction of synaptic weights of excited neurons allows us to increase their eigenvalues of discriminant functions with respect to input vectors. The correction is made in such a way that the output signal of the winner neuron is increased in case of the subsequent application of similar input vectors.

Thus, the basis of the SOM training procedure is the competition between neurons based on the value of the activation function in response to an incoming signal.

Assume that the inputs and synaptic weights are prenormalized, and the Euclidean measure is used as the distance:

$$d(x, w_i) = \|x - w_i\| = \sqrt{\sum_{j=1}^{N} (x - w_{ij})^2} \, . \tag{5.8}$$

Then, using the Euclidean measure, partitioning the space into zones of neuron dominance is equivalent to splitting it into Voronoi domains. If we use another measure (scalar product, measure relative to the norm of $L1$ – Manhattan – or a measure relative to the norm of $L\infty$), another division of neuron influence areas is formed.

With normalized input learning vectors, the weight vectors that strive for them are normalized automatically. It should be noted that when the weight vector is normalized, the Euclidean measure and the scalar product are equivalent to each other since

$$\|x - w_i\|^2 = \|x\|^2 + \|w_i\|^2 - 2x^T w_i \, . \tag{5.9}$$

Thus,

$$\min \|x - w_i\|^2 = \max(x^T w_i) \quad \text{at} \quad \|w\| = \text{const} . \tag{5.10}$$

The winner neuron is determined with the minimum distance to the presented input vector:

$$D\left(x(k), w_j^*(k)\right) = \min_j \|x(k) - w_j(k)\|^2 , \tag{5.11}$$

or for normalized vectors, that is the same as

$$D\left(x(k), w_j^*(k)\right) = \max_j x^T(k) w_j(k) = \max_j \cos\left(x(k), w_j(k)\right) . \tag{5.12}$$

It is obvious that

$$-1 \le \cos\left(x(k), w_j(k)\right) \le 1 \quad \text{and} \quad 0 \le \|x(k) - w_j(k)\|^2 \le 4 . \tag{5.13}$$

A topological neighbourhood $S_w(t)$ is formed around the winner neuron with a certain energy that decreases with time. The neuron-winner and all the neurons lying within its neighbourhood are subjected to adaptation, during which their weight vectors change in the direction of the vector x according to the Kohonen rule

$$w_i(t+1) = w_i(t) + \eta_i(t)(x - w_i(t)) \tag{5.14}$$

for $i \in S_w(t)$, where $\eta_i(t)$ is the coefficient of training of the i-th neuron on the neighbourhood $S_w(t)$ at time t.

The value $\eta_i(t)$ decreases with increasing distance between the i-th neuron and the winner.

Weights of neurons that are outside the neighbourhood of $S_w(t)$ do not change. The size of the neighbourhood and the training coefficients of neurons are functions whose values decrease with time. In [11] it is proved that adaptation according to the Kohonen rule is equivalent to the gradient method of training based on the minimization of the objective function,

$$E(w) = \frac{1}{2} S_i(x(t))[x_j(t) - w_{ij}(t)]^2 , \tag{5.15}$$

and $S_i(x(t))$ is a function of the definition of the neighbourhood, which varies during the learning process.

After the presentation of two different vectors x_1 and x_2, two network neurons are activated whose weights are closest to the coordinates of the corresponding vectors. These weights, denoted by w_1 and w_2, can be displayed in space as two points. Approximation of the vectors x_1 and x_2 causes a corresponding change in the arrangement of the vectors w_1 and w_2.

In the limit, the equality $w_1 = w_2$ is satisfied if and only if x_1 and x_2 coincide or are practically indistinguishable from each other. The network in which these conditions are met is called a topographic map or a Kohonen map.

Note that the tuning of the synaptic weights of the winner neuron can also occur in other algorithms: according to the WTA rule or with modified learning algorithms, giving better results than the WTA algorithm, for example, conscience WTA (CWTA), WTM or time-adaptive self-organizing map (TASOM) [1].

A common drawback of these algorithms' SOM is the presence of heuristic parameters and procedures for the solution of the problem of "dead" neurons increases the training time, with the impossibility of separating overlapping clusters. When network weights are initialized randomly, some of the neurons may be in the region of space in which there are no or a negligible amount of data. These neurons have little chance of winning and adapting their scales, so they remain dead. Thus, the input data will be interpreted by a smaller number of neurons, and the error in the interpretation of the data will increase. Therefore, an important problem is the activation of all neurons in the network, which can be done if the training algorithm provides for the calculation of the number of victories of each neuron and the training process is organized so as to give a chance to less active neurons to win.

Phase two: Additional terms and restrictions

We require that the desired self-organizing of the neuro-fuzzy network was possible under the following conditions:
1. High degree of self-organization of neurons;
2. An ability to selectively control individual connections between neurons to solve the problem of "dead" neurons;
3. High flexibility, simplicity of realization and temporal efficiency of algorithm;
4. Raising of dividing network properties in the case of overlapping clusters with output of the membership degree of the signal to each cluster.

Phase three: Architecture of Kohonen FCNNs

To ensure these conditions, the class of cellular neural networks (CNNs) was considered; they were introduced by Chua and Yang [24, 25] and became effectively used in computational models of image processing. In [23] the possibility of learning of Kohonen's neural nets by a cellular automaton (CA) that can significantly improve the quality and speed of learning was demonstrated; however, it did not resolve the issues dividing the network properties in the case of overlapping clusters.

We proposed a fuzzy cellular architecture of the Kohonen neural network (FCNN-SOM) that contains three layers (Figure 5.4):
1. Input (receptor) layer;
2. Layer of Kohonen neurons with lateral connections, trained by CA to identify centroids of overlapping clusters;

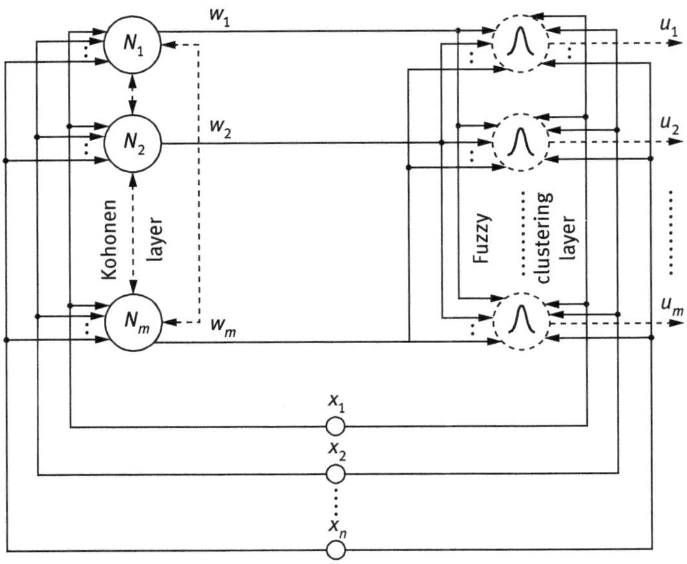

Fig. 5.4: Architecture of Kohonen FCNN-SOM

3. Additional (output) layer of fuzzy clustering calculating the degree of membership of the current vector in each cluster.

Consider the work of the FCNN-SOM in stages.

Phase four: Learning algorithm of Kohonen FCNNs

Source data – sample of observations, formed from N n-dimensional sign vectors $x(k)$, $k = 1, \dots, N$.

The goal of learning is the division of learning sample data into m clusters with some levels of membership $u_j(k)$ of the k-th sign vector in the j-th cluster ($j = 1, \dots, m$).

Step 1: Data preprocessing

Input data are centred and standardized for all ordinates so that all observations belong to hypercube $[-1; 1]^n$. Redefinition of the vector components is performed in accordance with the equation

$$x_{i,\text{new}} = \frac{x_i}{\sqrt{\sum_{i=1}^{N} x_i^2}} \,. \tag{5.16}$$

Centring can be done in two ways:

1. Relative to the average, calculated using the equation

$$m_i(k) = m_i(k-1) + \frac{1}{k}(x_i(k) - m_i(k-1))\,; \tag{5.17}$$

2. Relative to the median to make the centring procedure of the robust properties (protection against abnormal observations) by the recurrent equation

$$me_i(k) = me_i(k-1) + \eta_m \operatorname{sign}(x_i(k) - me_i(k-1))\,, \quad i = 1, 2, \ldots, n\,, \tag{5.18}$$

   where η_m is the parameter of the search step selected in accordance with the conditions of stochastic approximation of Dvoretzky [26].

Step 2: Initialization of initial state of neural network

Initialization of the initial state of the neural network is a self-adaptive deployment of network neurons with the formation of an ordered self-organizing Kohonen map. The main idea of this method is described in [23] and consists in the gradual retraction of inactive neurons in the normalized hypercube space of the training samples by active neurons.

Step 3: Training of Kohonen Layer by Moore's Cellular Automaton on Simply or Multiply SOM [23]

The cell of a CA is the neuron of the Kohonen layer, topologically bound in the Moore environment with eight (at the boundary – with three or five) neighbouring neurons of the two-dimensional self-organizing Kohonen map (Figure 5.5).

Thus, the CA is a set of Kohonen layer neurons interacting with its neighbours, placed in the continuous space of learning examples. The state of the CA cell is given by a multidimensional vector, whose components are the coordinates of the corresponding neuron in the space of the training samples. In accordance with the principle of the "nearest neighbour", network neurons are the centres of Voronoi multidimensional neighbourhoods, which, like a mosaic, share the entire space of training samples in the iteration of learning [23].

Any k-th neuron of the Kohonen map in the iteration of the epoch can change its multidimensional state vector in two cases:

1. The neuron becomes the winner;
2. The neuron is the neighbour of another winner neuron. Then the average total increment of the i-th component of the state vector of the k-th neuron with the Moore environment in one epoch will be

$$\Delta w_{ki} = \eta \frac{\sum_m \sum_{j=1}^{n_m} (x_{mi}^{(1)} - w_{ki})}{\sum_{m=k}^{k+8} n_m}\,, \tag{5.19}$$

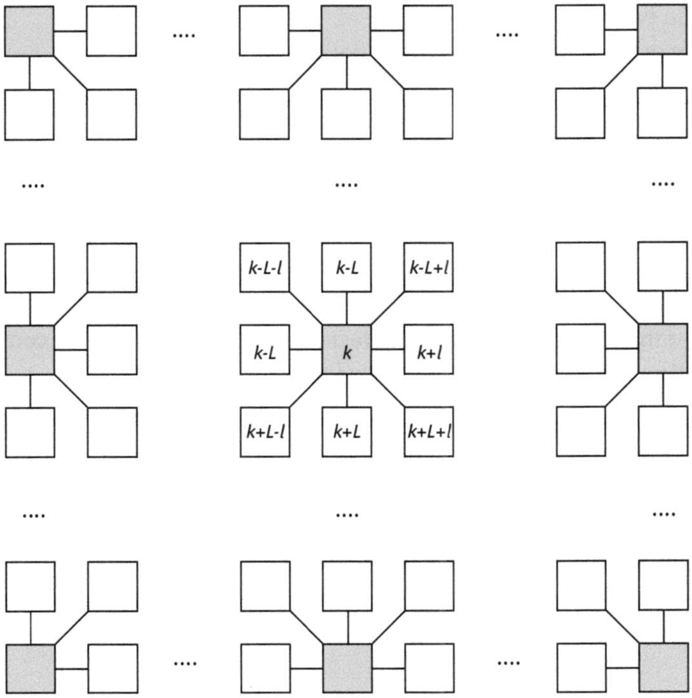

Fig. 5.5: Cellular topology of Kohonen map in Moore environment

where:

η is the learning rate;

k is the number of winner neuron;

$m = k - L - 1, k - L, k - L + 1, k - 1, k + 1, k + L - 1, k + L, k + L + 1$ are the numbers of the neurons from the environment of the k-th neuron;

L is the width of the rectangular Kohonen SOM;

n_m is the number of training samples that fall in the Voronoi neighbourhood of the m-th neuron;

$x_{mi}^{(j)}$ is the i-th coordinate of the j-th training sample from the Voronoi neighbourhood of the m-th neuron;

w_{ki} is the i-th coordinate of the k-th neuron.

Use the notation $S = \sum_{m=k}^{k+8} n_m$, then Equation (5.19) can be written

$$\Delta w_{ki} = \eta \frac{1}{S} \sum_m n_m (\overline{x}_{mi} - w_{ki}) , \tag{5.20}$$

where $\overline{x}_{mi}$ is the i-th coordinate of the centre of gravity of the Voronoi neighbourhood of the m-th neuron.

Equation (5.20) shows that in every learning epoch the k-th neuron is shifted to the relative value of η in the direction of the centre of gravity area of space of the training samples formed by the integration of the Voronoi neighbourhoods of the k-th winning neuron and its nearest neighbours. In particular, when $\eta = 1$, the neuron is shifted to the centre of gravity of the combined area [23].

Step 4 (new): Network training by algorithm of Bezdek's fuzzy clustering instead of WTA algorithm when power of lateral connections in Kohonen layer is disabled
Among the known algorithms of fuzzy clustering, the most reasonable from a mathematical point of view, we noted the approach based on target functions [16] that solves the problem of optimization under certain assumptions, for example, a probabilistic algorithm based on minimizing the following criterion (objective function):

$$E(u_j, w_j) = \sum_{k=1}^{N} \sum_{j=1}^{m} u_j^{\beta}(k) \|x(k) - w_j\|^2 \tag{5.21}$$

with conditions

$$\sum_{j=1}^{m} u_j(k) = 1\,, \quad 0 \le \sum_{k=1}^{N} u_j(k) \le N\,, \quad u_j(k) \in [0, 1]\,, \tag{5.22}$$

where:

$\beta \ge 0$ is a parameter of fuzzification (fuzzifier), which determines the fuzziness of borderlines between clusters, $k = 1, \ldots, N$;

$u_j(k)$ is the level of membership of the k-th feature vector in a j-th cluster ($j = 1, \ldots . m$).

The result of clustering is an $(N \times m)$ matrix $U = \{u_j(k)\}$, called a matrix of fuzzy partitioning. The elements of the matrix U can be considered as the probability of the hypotheses of membership of vectors in specific clusters. Inserting the Lagrangian objective function:

$$L(u_j(k), w_j, \lambda(k)) = \sum_{k=1}^{N} \sum_{j=1}^{m} u_j^{\beta}(k) \|x(k) - w_j\|^2 + \sum_{k=1}^{N} \lambda(k) \left(\sum_{j=1}^{m} u_j(k) - 1 \right) \tag{5.23}$$

where $\lambda(k)$ is an undetermined Lagrange multiplier, and solving the system of equations of Karush-Kun-Tucker

$$\begin{cases} \dfrac{\partial L(u_j(k), w_j, \lambda(k))}{\partial u_j(k)} = 0\,, \\ \nabla_{w_j} L(u_j(k), w_j, \lambda(k)) = 0\,, \\ \dfrac{\partial L(u_j(k), w_j, \lambda(k))}{\lambda(k)} = 0 \end{cases} \tag{5.24}$$

we obtain the required solution in the form

$$u_j(k) = \frac{(\|x(k) - w_j\|^2)^{\frac{1}{1-\beta}}}{\sum_{l=1}^{m}(\|x(k) - w_j\|^2)^{\frac{1}{1-\beta}}}, \quad w_j = \frac{\sum_{k=1}^{N} u_j^{\beta}(k)x(k)}{\sum_{k=1}^{N} u_j^{\beta}(k)},$$

$$\lambda(k) = -\left(\left(\sum_{l=1}^{m} \beta\|x(k) - w_j\|^2\right)^{\frac{1}{1-\beta}}\right)^{1-\beta}, \tag{5.25}$$

coinciding at $\beta \to 1$ with an algorithm of a hard K-means (HKM) [27] and at $\beta = 2$ with the algorithm of a fuzzy C-means of Bezdek (FCM) [28]:

$$u_j(k) = \frac{\|x(k) - w_j\|^{-2}}{\sum_{l=1}^{m} \|x(k) - w_l\|^{-2}}, \quad w_j = \frac{\sum_{k=1}^{N} u_j^2(k)x(k)}{\sum_{k=1}^{N} u_j^2(k)}. \tag{5.26}$$

We rewrite Equation (5.26) in the form

$$u_j(k) = \frac{1}{1 + \|x(k) - w_j\|^2 \sum_{\substack{l=1 \\ l\neq j}}^{m} \|x(k) - w_l\|^{-2}} = \frac{1}{1 + \frac{\left\|\|x(k)-w_j\|^2\right\|}{\sigma_j^2(k)}}. \tag{5.27}$$

We obtain an expression that specifies the bell-shaped membership function with the centre in point $w_j(k)$ and setting

$$0 \leq \sigma_j^2 = \frac{1}{\sum_{\substack{l=1 \\ l\neq j}}^{m} \|x(k) - w_l(k)\|^{-2}} \leq \frac{4}{m-1}, \tag{5.28}$$

where m is the number of clusters.

A bell-shaped membership function is characterized by the smoothness, symmetry and simplicity of the writing, and this is the most commonly used function to describe fuzzy sets (Figure 5.6).

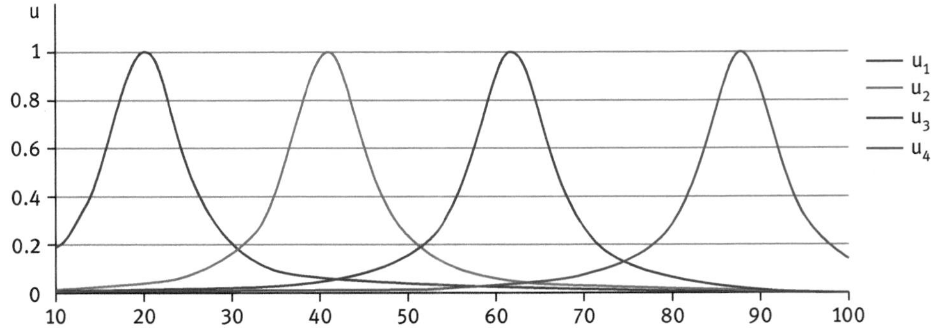

Fig. 5.6: Curve of bell-shaped membership functions

Thus, the proposed method not only calculates the parameters of the centres and the width but, unlike [29, 30], automatically resolves the issue of the type of membership function used for data fuzzification. Assumes that the whole sample that is subject to clustering is predefined and cannot change during processing.

5.5 Development of a model of a generalized deep neuro-fuzzy classifier

The architecture of a generalized deep neuro-fuzzy classifier combining FCNN-SOM and RBFN is shown in Figure 5.7.

The concept of this model corresponds to the Pospelov classification [1], which is based on two forms of knowledge representation, extensional (the formation of clusters of generalized images) and intensional (identifying integral characteristics in the selected clusters of classes).

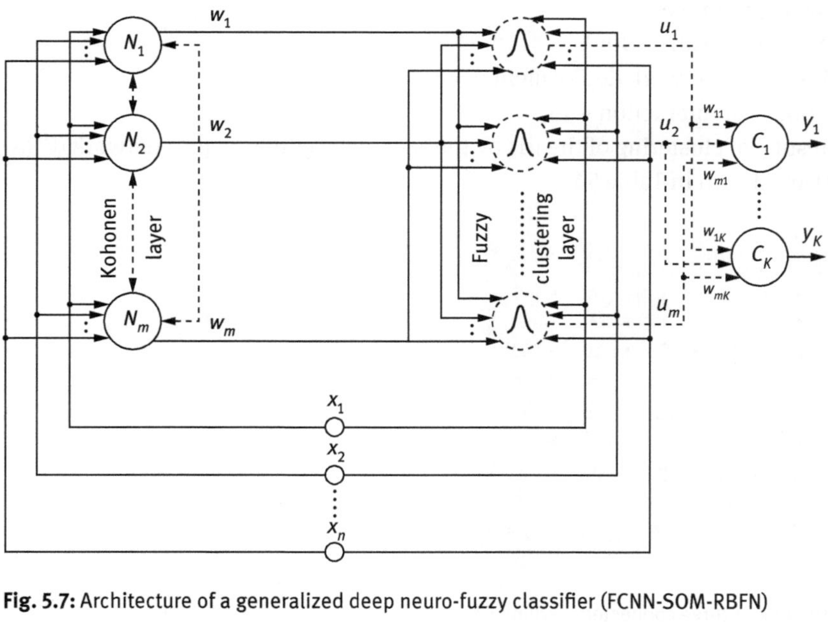

Fig. 5.7: Architecture of a generalized deep neuro-fuzzy classifier (FCNN-SOM-RBFN)

5.6 Experimental results and discussion

The experimental study was conducted in stages on a well-known example of the classification of iris flowers (Fisher problem), where there are three classes of iris flower: Setosa (class 1), Versicolor (class 2), and Virginica (class 3). For each iris flower four independent parameters were measured: length and width of sepal and length and width of petal.

Step 1: Normalized four-dimensional training sample data contained 50 specimens of iris flowers of classes 1, 2 and 3, and the samples of class 1 were linearly separable from the samples of classes 2 and 3.

Step 2: The initial coordinates of the neurons that are set while initializing the network:

Neuron 0 – (–0.55, 0.08, –0.78, –0.83);
Neuron 3 – (0.20, –0.30, 0.20, 0.5);
Neuron 15 – (0.15, –0.30, 0.50, 0.68);

the other neurons were placed on a hypersphere of initial conditions at coordinates (0, 0, 9.33, 3.60). Note that the coordinates of neurons 0, 3 and 15 were selected coincident with the coordinates of one of the training samples of classes 1, 2 and 3 respectively, which is necessary for successful activation of all neurons of the network in the deployment phase.

Step 3: The Kohonen map was trained by three connected CAs (Figure 5.8) with the following parameters:

Number of runs – 30;
Epochs of network deployment – 5;
Epochs of interaction – 25.

The results of clustering of the iris by the Kohonen cellular neural network (CNN-SOM) are shown in Table 5.1.

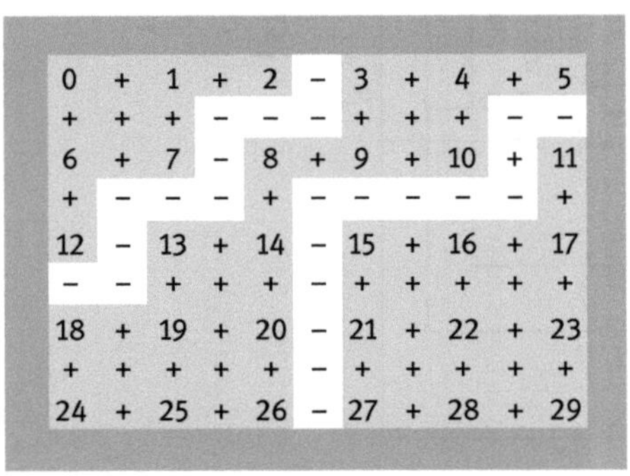

Fig. 5.8: Structure of three connected learning CAs

Tab. 5.1: Dividing neurons into clusters

Cluster	Neurons of cluster
1	0, 1, 2, 6, 7, 12
2	3, 4, 5, 10, 18, 19, 20, 24, 26
3	11, 15, 16, 17, 21, 22, 23, 27, 28, 29
4	8, 9, 13, 14, 25

An additional cluster 2/3 is formed by network neurons located on the border between linearly inseparable clusters 2 and 3. For a more accurate separation of clusters 2 and 3, an additional stage of fuzzy clustering was conducted (step 4).

Step 4: The lateral connections in the Kohonen layer are disabled, and centroids calculated in the previous stage wj(k) were applied to the layer of membership degree calculation. The network is additionally trained by the algorithm of Bezdek fuzzy clustering with the following parameters:

Number of clusters – 3;

Maximum number of iterations – 100;

Minimum value of improvement in objective function during one iteration – 0.000001.

A graphical illustration of the initial classification of iris data and the results of fuzzy clustering by FCNN-SOM are shown in Figure 5.9.

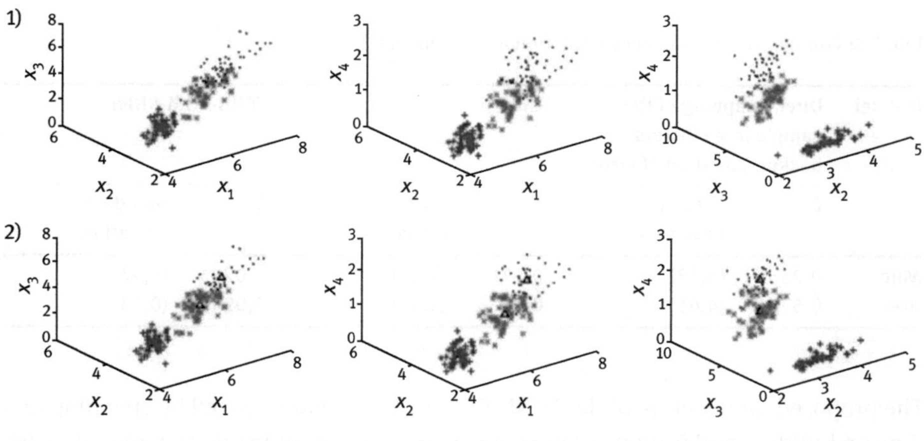

Fig. 5.9: Comparative results of clustering on Iris data set: (1) original data and classification of irises; (2) results of fuzzy clustering by FCNN-SOM

Comparative computational experiments on FCNN-SOM-RBFN-based data classification versus the method of direct mapping of samples to clusters with training by an error propagation inverse algorithm and fuzzy perceptron (FMLP) were carried out on two sets of test data (Table 5.2) from the UCI repository:

Tab. 5.2: Characteristics of test sets

N	Test set	Number of examples (*N*)	Number of variables (*n*)	Number of classes (*K*)	Error (ME)
1.	Wine	178	13	3	59.74
2.	Liver	645	6	2	57.97

1. Wine with probabilistic separability of classes. The data set originates from a chemical analysis of wines from grapes grown in the same region but belonging to three different varieties. The analysis identified the percentage of 13 constituent wines present in each of the 3 varieties. The task is to determine which of the three varieties a wine with given characteristics belonged to. The total amount of data comprised 178 samples, of which 142 were trained and the remaining 36 were used for testing.
2. Liver (liver disease) for intensely intersecting classes. The sample contained data of medical blood tests of 345 patients. Signs represent the test results for six indicators. The purpose of the task was to identify patients with liver disease.

Table 5.3 shows the average error rates *E* of classification on test data for ten multiple cross-validations; the standard deviations of errors are presented in parentheses.

Tab. 5.3: Comparative results of computational experiments

Test set	Direct mapping of the sample into clusters, backpropagation of error		FMLP		FCNN-SOM-RBFN	
	E	Standard deviation	*E*	Standard deviation	*E*	Standard deviation
Wine	0,2537	(6,43)	0,1212	(2,19)	0,0285	(1,47)
Liver	0,5341	(4,65)	0,1627	(1,84)	0,0943	(0,13)

The proposed architecture of the FCNN-SOM-RBFN is characterized by spending less time on building and training of the network with a minimum error compared to displaying the sample in clusters with learning by backpropagation of error and fuzzy perceptron (FMLP). At the same time, it provides higher average accuracy of models and a significantly higher level of generalization in comparison with known methods which are provided by the good localization of data in a layer of the FCNN-SOM input space in the form of generalized images and automatic tuning of parameters of the membership function, so the question of number of clusters is automatically resolved. This allow to simplify of network by reducing its structural and parametric complexity, and improves the network's dividing capacity in the case of intensely intersecting classes (Liver).

5.7 Practical results of modelling

In this section we describe the methodology of a multicriteria neuro-fuzzy network analysis of global technical and economic development (TED) for a set of quantitative and qualitative indicators, which characterizes the modern technological order and examines issues of technological and social measurement of the transition process to the sixth technological order. For a fuzzy separation of overlapping clusters, we use a new hybrid architecture – FCNN-SOM. The modified neuro-fuzzy network model of the trajectory of global TED, distinguished by a fairly simple but powerful strategy for improving clustering accuracy interpretability, assessing the level of TED and the characteristics of the TED of individual countries and predicting future values of the multidimensional process, is based on it. The trajectory of the actual TED of the leading countries of the corresponding technological order can be considered as a leadership trajectory of TED. Intercountry comparison of model characteristics on the leadership and national TED trajectories make it possible to obtain predictive estimates not only of speed but also of the level of TED of each country. The experimental results we have obtained show the effectiveness of the new architecture and clustering algorithms.

Methodological approaches to the study of economic cycle phases

The relevance of this research topic is due to the fact that the problem of long-term fluctuations in economic dynamics and methods of global modelling of these processes was of central importance in global studies of Russian and foreign scientists in the twentieth century and has especially attracted the attention of modern scientists in this transition period from the fifth to the sixth technological order. In the last decade, based on the scientific schools emerging from the work of N. D. Kondratiev and J. Schumpeter, a new paradigm of economic science has formed, and their representatives have merged into the international research network known as GLOBELIX.

Cyclicity is the universal form of movement of national economies and the world economy as a single entity. It expresses the uneven functioning of various elements of national economies, the replacement of revolutionary and evolutionary stages of their development, and economic progress. Finally, cyclicality is the most important factor of economic dynamics, one of the determinants of macroeconomic equilibrium. It is often extremely difficult to single out individual cycles because of the difficult, mutually intersecting trends of the various components of cycles. The most characteristic feature of cyclicity – movement – occurs not in a circle but in a spiral. Therefore, cyclicity is a form of progressive development. Each cycle has its own phases, its own duration. Characteristics of phases are unique in their specific indicators. They are original both in historical and in regional aspects.

The emergence of a market economy in a number of countries and their inclusion in the global economic space dispelled the myth of overcoming trends in cyclical dynamics and required new intellectual interdisciplinary approaches to retrospective analysis and forecasts of large cycles of conjuncture (long waves of economic dynamics) introduced by N. D. Kondratieff [31]. The task is complicated by the action in national economies of a number of factors that alter the course of economic cycles, the lack of sufficiently reliable and long-term time series. Because of its unique features, each national economy has its own rational trajectory of future fuel and energy resources. Its planning, possible only in the form of approximate scenarios, should be based on global TED trends taking into account the aforementioned national characteristics. The most important of these is a country's position in the intercountry hierarchy of TED. To conduct such an assessment, the model of the leadership fuel–energy trajectory, which establishes a reference frame and scale common to all countries, is relevant and serves as a basis for measuring the technical development of national economies.

Thus, in our opinion, the correct assessment of the technical development of a national economy, its position in the world technical and economic development, is related to coming up with an adequate method of measuring the level and rates of TED relative to the leadership trajectory.

Due to the patterns of production of social capital, the life cycle of the technological order in a market economy is reflected in the specific form of the long wave (LW) of economic conjuncture. The so-called long waves, or Kondratieff waves, are the subject of study of a special direction of economic research – the analysis of long-term processes of social production. In the modern periodization, there are five long waves (Figure 5.10).

In empirical studies of long waves, a fundamental one-pointedness of the technological changes taking place in different countries, a similarity of national TED tra-

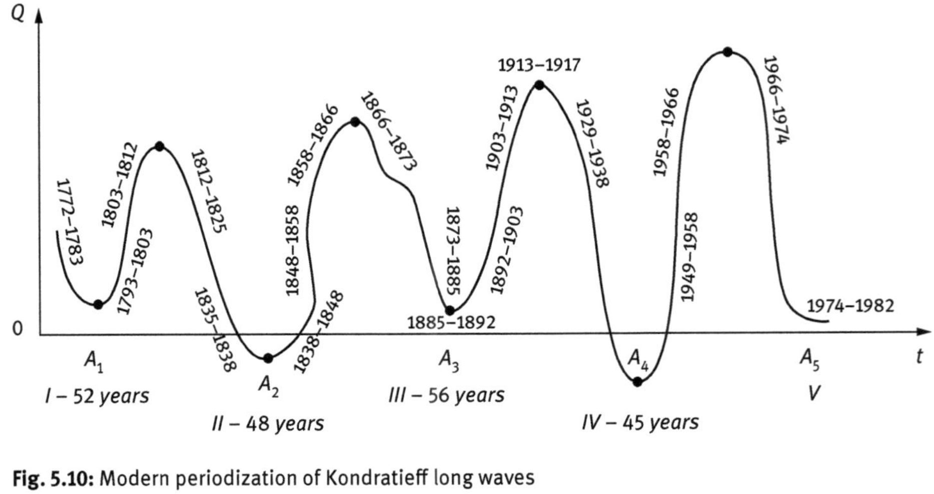

Fig. 5.10: Modern periodization of Kondratieff long waves

jectories and also a tendency to synchronize macroeconomic fluctuations and technological changes have been established.

In particular, the same shape of fuel and energy resource trajectories in countries with both market and directively managed economies had been identified in the structure of energy consumption, in metallurgy and in extractive industries, in the dynamics of transport infrastructure and in other sectors of the economy. The unidirectional nature of energy resources in different countries, as well as the formation of a single rhythm of the world economic system, is due to the establishment of a global market and the rapid expansion of international economic ties since the industrial revolution. The experience of macroeconomic research tells about not only the possibility but also the fruitfulness of using cross-country comparisons to obtain both qualitative and fairly accurate quantitative conclusions, including those related to making forecasts.

Many researchers have noted that global technological changes are generated by countries that are leaders in the life cycle of the corresponding technological orders (TO) [32]. Technological order is a set of related industries with a single technical level and developing synchronously. The change in the economy of the dominant technological orders determines the uneven progress of scientific and technological progress [32]. Although the technological shifts that make up the content of the life cycle of each technological order take place on the scale of the world market, the economic structure of the leading countries most fully projects the structure of the corresponding TOs, and the dynamics of their TED projects the evolution of these technological orders. Therefore, the trajectory of the actual TED of the leading countries of the corresponding technological order can be considered a leadership trajectory of TED.

There are several types of economic cycles known as waves. They are difficult to distinguish because of the multiplicity of their indicators and because of the temporary blurring of the boundaries between them. So-called long waves (cycles) have a length of 40–60 years. The development of the theory of long waves began in 1847, when the Englishman H. Clark drew attention to the 54-year gap between the crises of 1793 and 1847. He suggested that it was not by chance that the gap was objectively conditioned. A significant contribution to the development of the theory of long waves was made by his compatriot V. Jevons, who first drew statistics on price fluctuations to explain a phenomenon new to science.

K. Marx made a significant contribution to the theory of cyclicity. He paid attention to the study of short waves, which in the economic literature came to be known as periodic cycles or periodic crises of overproduction. Each cycle, according to Marx, consists of four phases: crisis, depression, revival and recovery, which completely agrees with the theory of cyclicity [31].

The mention of long-term fluctuations can be found in the studies of M. Tugan–Baranovsky. The theory of cyclicity was reflected in the works of the Russian scientist A. Gelfand (Parvus), who attempted to prove that cyclicity was immanent in capitalism. The original statistical treatment of the material is contained in the works of Dutch scientists J. Gelderen and S. Wolf. The novelty of their studies was that they considered technical progress as a factor of cyclicity and analysed the terms of the functioning of the transportation infrastructure.

The understanding of the cyclical nature of economic development became especially widespread in the late nineteenth and early twentieth centuries, when scientists from many countries began paying attention to the dynamics of certain economic indicators. This justified classifying the theories of economic cycles, which was most fully presented by W. Mitchell in his work *Economic cycles*. The American economist distinguished the following theories of economic cycles:

1. Theories that reduce economic cycles to natural-physical processes. These theories explained the cyclicity of economic life by cycles of solar radiation, changes in the position of Venus relative to the Earth, and meteorological conditions.
2. Theories that reduce economic cycles to psychological causes that create either a favourable or unfavourable environment for economic activity. According to some economists, fluctuations in the mass sentiments of people precede fluctuations in wholesale prices and influence decision-making in the economic sphere. The basis of these theories is a fact noted in public life, indicating the periodicity of changes in society of motivational structures, the creative activity of the population and social moods. Cyclical changes in the socio-psychological variable, encompassing the "degree of activity" of the population, its enterprising, "optimistic outlook" and other individual and socio-psychological aspirations of people, determine long-term cycles of economic development.
3. Theories that reduce economic cycles to institutional processes. The essence of cyclicity, from the standpoint of these theories, is that economic cycles arise from changes in economic institutions. The greatest influence on the formation of the cyclical nature of economic development is exerted by the functioning of existing economic institutions.
4. Other theories proceed from the assertion that cycles arise because of imbalances in the processes of total production and consumption of goods. Favourable trade and industrial conditions lead to a rapid growth of industrial equipment, later to an increase in output and, ultimately, to a fall in the marginal prices of consumer goods. A consequence of this is depression, during which the growth in the amount of industrial equipment and products is interrupted. Finally, the marginal prices for consumer goods rise again, and a new period of recovery begins.

The most profound development of the concept of economic cycles is primarily connected with the works of the national economist N. D. Kondratieff [31]. He put forward a hypothesis about the mechanism of long cycles in the economy ("large cycles of conjuncture" in his terminology), linking them not only with the price dynamics, but also with the process of capital accumulation, the growth rates of production and the dynamics of innovation, with the constant evolution of the national economy. The motion of long-term oscillations according to Kondratieff proceeds according to the following principle. Before the start of a large cycle, a sufficient amount of free funds accumulates – the growth of bank reserves, which allows creditors to reduce the interest rate. The percentage of long-term credits is low. During the previous period of decline, a significant amount of technical innovations (inventions) accumulated. Under these conditions, capital investment in large facilities begins to increase, which causes dramatic changes in the conditions of production (technical opportunities for earning profits are generated), while production becomes profitable, and then there is an upswing in a large cycle of economic conjuncture.

The main modern theories of long waves:

1. Innovation theory (J. Schumpeter). The essence of this theory is that there are people who want to earn entrepreneurial profits. They invest in new technologies and industries; then come their followers. This is a period of recovery. Gradually, the market is saturated, profit is reduced, and people withdraw their funds from production – this is a recession.
2. The theory of overaccumulation in the capital sector (D. Forrester). The capital sector, which produces the means of production, provides machines and equipment not only for the industries that produce consumer goods but also for themselves. The growth of consumption causes even faster growth in the means of production, that is, an accelerator acts between the two branches. The magnitude of this accelerator in real life is much greater than that required for equilibrium motion. All these factors contribute to overaccumulation in the capital sector. Orders first increase sharply and then abruptly contract. This is sufficient for the appearance of long-term oscillations.
3. Theories related to labour (K. Freeman). This group of theories is based on a consideration of the theory of long waves from the point of view of the laws of the labour force. The mechanism is the introduction of new technologies that bring to life new industries. In the early stages of the application of new technologies, the demand for labour is limited. This is due to the fact that the volumes of new production are still small and require, not a mass, but a particularly skilled labour force. Gradually, production increases, and the demand for labour begins to increase. This growth continues to saturate demand for both labour and related goods. In parallel, wages rise and costs increase. There is a need for labour-saving innovations. There is an outflow of labour and a decline in wages and general demand, that is, a decline in the economy.
4. Price theories (W. W. Rostow). In the theories examined, the prices of goods either were not considered at all or played the role of indicators of the processes taking place in the sphere of production. However, the pricing process and price dynamics are directly related to the explanation of the long-term cycle and its turning points. Changes in the demand and supply of raw materials and food products, and, consequently, their prices, affect the innovation activity that determines the sequence of the leading industries and itself depends on them.

In the structure of a cycle, the higher and lower points of activity and the phases of recession and growth lie between them. The total cycle time is measured by the time between two neighbouring higher or two adjacent lower activity points. Accordingly, the duration of a decline is the time between the highest and the next lowest activity points, and the rise is the opposite. The National Bureau of Economic Research stated that in the development of the US economy from 1854 to 1991, there were 31 cycles. The average time between the two highest points was 53 months, of which 18 months accounted for the recession and 35 months the growth period.

A more detailed analysis divides the economic cycle into four phases (Figure 5.11).

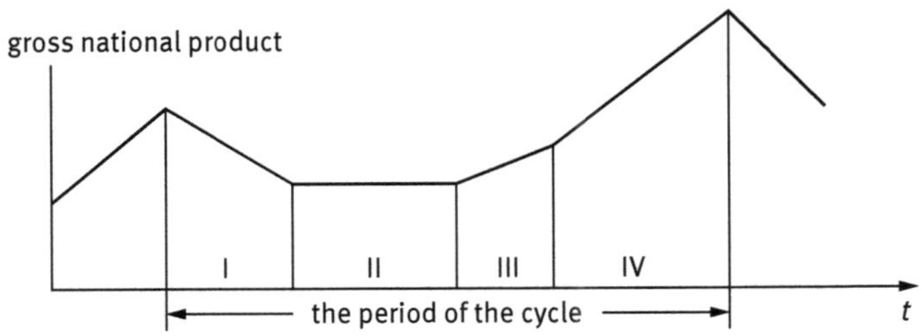

Fig. 5.11: Modern periodization of Kondratieff long waves

Phase one: Crisis phase (recession)

The main manifestation of a crisis is a fall in production volumes and a reduction in the size of the gross national product. Accordingly, the production capacities of enterprises are not fully exploited, profit rates are falling, share prices are falling, employment is falling, wages are falling, living standards are falling and poverty is increasing. As a result, aggregate demand decreases, and in response, production and, hence, supply fall further. In general, this phase is characterized by an excess of aggregate supply or aggregate demand. To ensure macroeconomic balance, it becomes necessary to reduce production.

Over time, the crisis can last from several months to several years, as happened during the Great Crisis of 1929–1933.

Phase two: Depression phase

This phase is characterized by the fact that following the liquidation during the crisis of commodity stocks, the production decline ceases, but there is still no growth. Accordingly, employment is at a low level, but unemployment has already bottomed, albeit at a low level, wages and profits of enterprises have stabilized, and business activity is not high. The depression can last from a few months to several years. For example, the Great Depression that began in 1933 after the Great Crisis lasted until 1938, that is, practically until the war itself.

Phase three: Recovery phase

The name of the phase speaks for itself. It is characterized by a revival of the economy, some gross national product growth occurs, the demand for labour and loan capital is increasing. The most important thing is that the investment activity of enterprises is intensified. Usually this phase does not last long; it quickly passes into the next phase.

Phase four: Lifting phase

This phase is also called a boom, since it is characterized by fairly rapid economic growth. There is an increase in employment, unemployment is receding, and some industries may experience a labour shortage. Wages are rising, aggregate demand is growing, sales volume is growing, business profits are increasing. The population and enterprises have free money, their supply on money markets increases, so interest rates do not increase and sometimes start to decline. Credit becomes comparatively inexpensive. The share price of many enterprises has a tendency to grow.

The problems of the theory of economic cycles cause the application of complex dynamic models using differential equations [33].

To characterize the economic situation, a number of economic indicators are used (gross national product, unemployment rate, personal income, industrial output, price level and many others). Depending on how the value of economic parameters varies during the cycle, they are divided into pro-cyclical, countercyclical and acyclical. The pro-cyclical parameters in the during growth periods increase, and in the arecession phase they decrease (e.g., loading of production capacities, aggregates of the money supply, general price level and corporate profits).

Countercyclical parameters are indicators whose value decreases during a recession and decreases during growth periods (e.g., unemployment rate, number of bankruptcies and stocks of finished products).

Acyclical parameters are those whose dynamics do not coincide with the phases of the economic cycle (e.g., the volume of exports).

According to the classification of the US National Bureau of Economic Research there are three types of parameters on the basis of synchronization – forward-looking, laggards and coinciding. Forward-looking parameters reach a maximum or a minimum before approaching the peak or the lowest point (these are changes in stocks, money supply, etc.). The laggards parameters reach a maximum or a minimum after a peak or a low point (the number of unemployed, the specific costs of wages, etc.). The coinciding parameters change in accordance with the fluctuations in economic activity (gross national product, inflation, industrial production, etc.).

Currently, statisticians and economists are not able to give accurate forecasts of the economic situation but can only determine its general trend. Firstly, it is difficult to take into account all factors, especially during a period of economic instability and political turmoil. Secondly, the international environment has a significant impact on the national economy. Thirdly, even correctly determining the trend, it is difficult to predict the exact dates phases will be reached and change economic policy in time. Finally, the actions of entrepreneurs can exacerbate undesirable deviations in conjunctures.

Technological and social dimension of TED

Prior to the results of the comparative analysis of TED, two factors should be distinguished. On the one hand, the objects and territories of states, regardless of their size and role in overall economic development, are not self-sufficient economic entities; in addition, despite the existence of internal trends and interrelations, states cannot plan their development based only on their own dynamics and ratios. On the other hand, although countries united by the international division of labour are developing along general lines of technical and economic evolution, they differ significantly in terms of the absolute level of TED indicators (measured in relative units – per inhabitant or unit of national income), even if they are on the same level of TED. This is explained by historical, scientific, educational, cultural, psychological, natural and climatic and other features of each country, which are reflected in its economic structure [33]. Therefore, to construct an adequate model of the fuel–energy trajectory in the group of parameters studied, we introduced a qualitative growth component, the index of productivity of primary resources, which is measured as the ratio of gross national product to the value of primary raw materials consumed by the economy [33]. Primary raw material resources are a set of such primary organic and inorganic resources that are used on a large scale both for consumption by the general population and for further processing of material production (food, construction materials, fuel and energy resources).

Leading countries that have the lowest energy and material usage intensity for the main types of resources nevertheless consume 1.5 to 2.5 times more primary raw materials per capita than less developed countries such as Russia, Brazil, India and China.

Productivity as an indicator of the effectiveness of resource use reflects progress in science and technology. Even in a period of recession (1991–1998), Russia was experiencing quite intensive processes improving the quality of products and services (Figure 5.12).

This was reflected in the fact that the productivity of primary resources increased by an average of 2.3% per year. Taking into account the fact that during this period the dynamics of the quality component of growth for the USA decreased substantially (to 0.95% per annum compared to 2.3% a year on average over the previous 30 years), the index of improving exchange ratios for the Russian economy was 1.016.

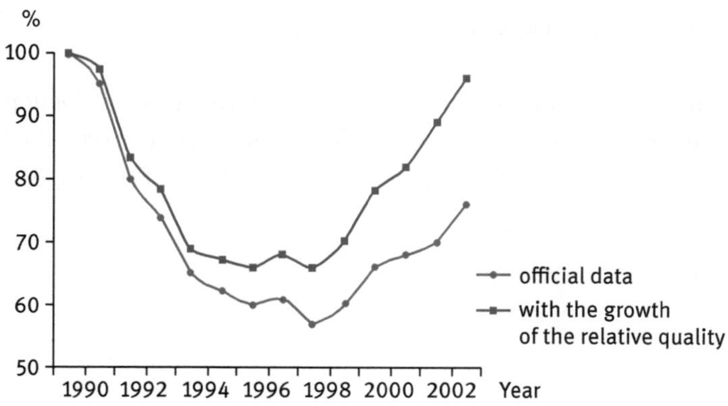

Fig. 5.12: Dynamics of Russia's gross national product in 1990–2003 at level of 1990

Since 1998, during the recovery period, the dynamics of the quality component of growth in Russia increased to almost 4% per year. However, as the qualitative growth component for the USA also increased during that period (up to 2.4% per year), the index of change in the ratios of exchange changed insignificantly – to 1.017.

Therefore, the gap in the economic development of Russia and the USA can be completely overcome if the primary resources productivity index (science and technology) in Russia grows faster. When modelling the TED trajectory, it is necessary to take into account that innovations reveal and define the essence of the technical process at the present stage of history. Selected quantitative and qualitative indicators of TED were supplemented by indicators of the country's innovative potential, divided into four groups:

1. Indicators characterizing the financial component of the innovation potential (e.g., the share of investment relative to gross national product, the share of investments in their total volume);
2. Indicators characterizing the material component of the innovation sphere (number of organizations associated with scientific activity, the value of their fixed assets, the relative share of scientific organizations).
3. Indicators characterizing the personnel component of the innovation potential (number of workers associated with innovation, their average salary, average age, relative share).
4. Indicators characterizing the resulting component of the innovation potential as a factor of economic growth and showing its ability to produce an effect (sales volumes, costs and profits associated with innovation, the number of created, exported and imported technologies, filed patent applications and issued patents, their relative share).

Neuro-fuzzy model of world technical and economic portrait

As a TED trajectory, a multidimensional time series is modelled, which is a collection of several one-dimensional time series, each of which describes the change over time of any of the listed characteristics characterizing the object of research (Figure 5.13).

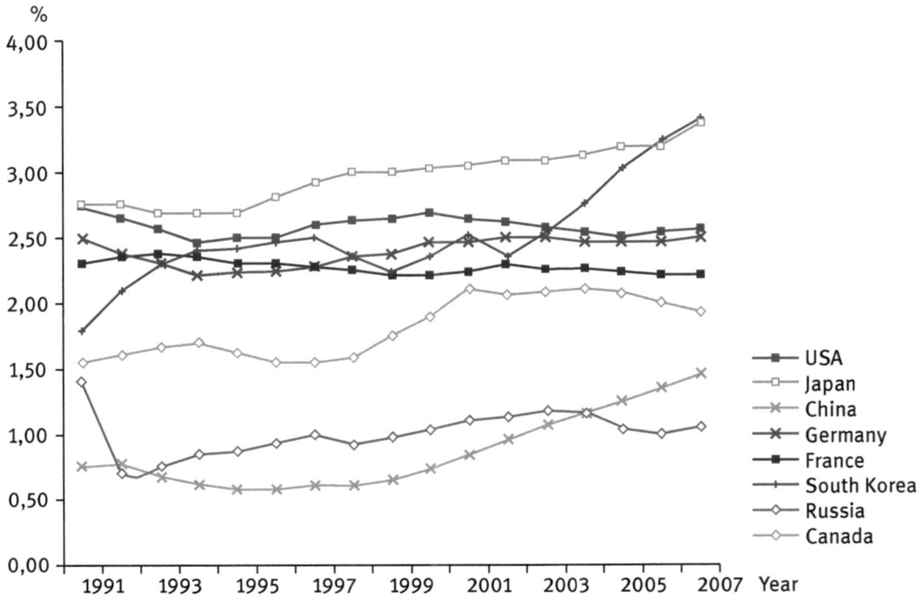

Fig. 5.13: One-dimensional time series by country (indicator – percentage of expenditure on research and development in gross national product dynamics)

For each year t, a particular country is represented by a point in the multidimensional space of quantitative and qualitative characteristics.

In our opinion, statistical processing of data for the description of the TED trajectory cannot be recognized as a reliable and adequate tool for performing such weakly structured tasks. It does not allow you to establish the cause-effect relationships of the parameters of the predicted system as a whole. In addition, the construction for each country of its generalized TED curve by linear contraction of the input vector space and the intercountry comparative analysis is not entirely incorrect due to, firstly, the "fluid" non-equilibrium nature of TED, with waves of falling and take-off (Kondratieff waves), and, secondly, which is very important, the input parameters for trajectory construction, in our opinion, should not be rigidly specified – they depend both on the national system of accounts in each country and on historical, cultural-psychological, natural-climatic and other features of the country at the time of observation t,

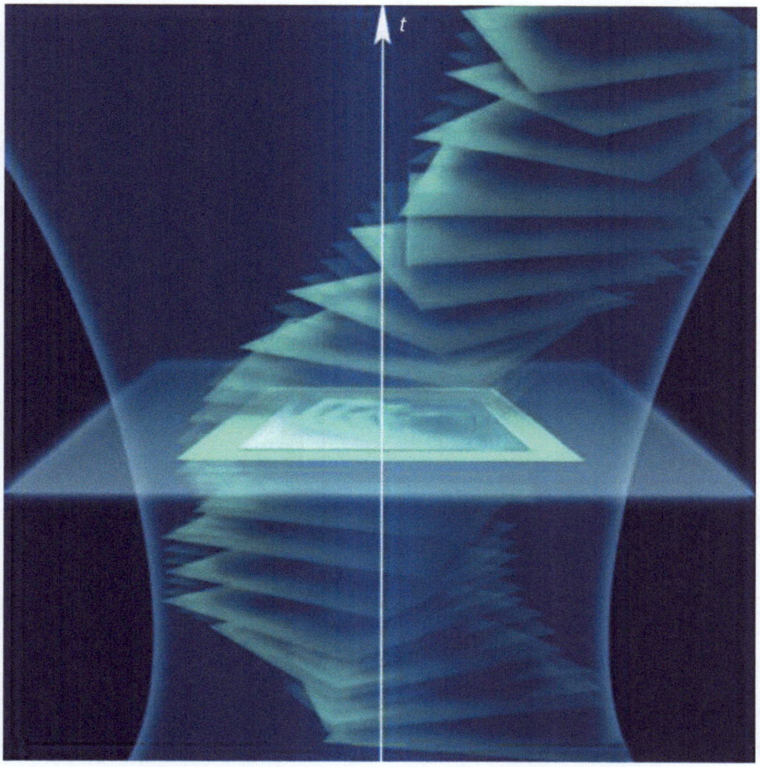

Fig. 5.14: Model of TED trajectory as time series of Kohonen SOM in multidimensional feature space

while the significance of the same parameters for the construction of the model will be different for different technological structures and phases of cycles.

To improve the modelling quality, in [33] we proposed a complex neural network analysis of quantitative and qualitative indicators, which in the training process are ranked by weight coefficients by the level of significance for solving the given problem. Based on the available data samples, a time series of Kohonen self-organizing maps (SOMs) was constructed, reflecting the dynamics of global TED (Figure 5.14).

This model makes it possible to simplify a multidimensional structure; it can be considered one of the methods of projecting a multidimensional space into a space with a lower dimension.

Figure 5.15 shows a neural network model of the world technical and economic portrait reflecting the current hierarchy of countries in 2010.

Each country in terms of its TED level in the current year is displayed by a specific cell on the map. Cells with the same coordinates contain countries with a similar TED state. The further away on the map the coordinates of countries are, the more their technical and economic portraits differ from each other.

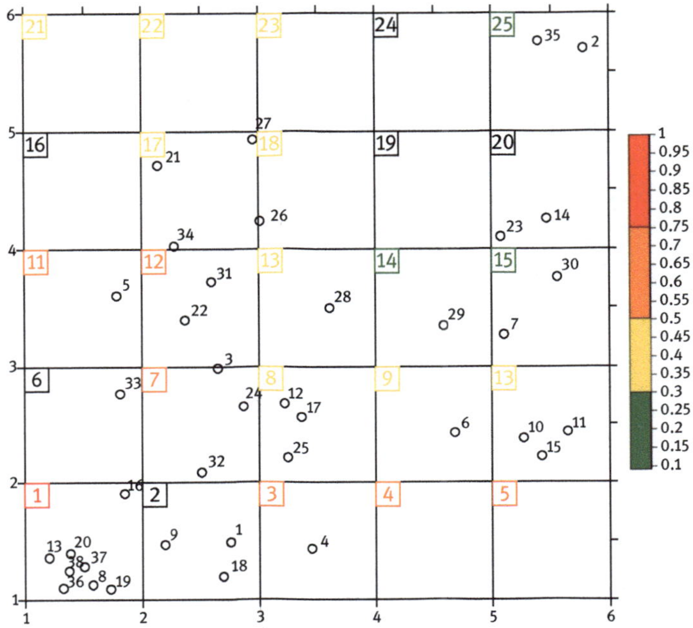

Fig. 5.15: SOM model of world technical and economic portrait in 2010

Based on the results of the training, four clusters of countries with similar levels of TED relative to the reference value were identified (Table 5.4).

Thus, network neurons that are located at the boundaries between linearly inseparable, intersecting clusters form additional clusters 1/2, 2/3 and 3/4. For more accurate separation of clusters, we used FCNN-SOM – Figure 5.16) and Table 5.5 show the results of fuzzy clustering.

Each country in the i-th year of observation is characterized by a set of parameters (Y_i, A_i, B_i, C_i), where:

Y_i – level of fuel and energy resources of country in i-th year;

A_i – actual distance (number of years elapsed since reference level of TED parameter corresponded to level of country in question in i-th year of observation);

B_i – prospective distance (number of years that will be required by a given country, starting from i-th year, to reach standard level of technical development in i-th year of observation);

C_i – conditional distance (number of years that this country needs to enter reference trajectory).

Tab. 5.4: Dividing of neurons into clusters

Cluster	Cluster neurons	Group of countries	Colour on map
1	1	Leaders of TED	Red
2	3,4,5,7,11,12	Developed countries	Orange
3	8,9,10,13,17,18,21,22,23	Countries with a moderate level of TED	Yellow
4	14,15,25	Developing countries	Green
1/2	2,6	?	Black
2/3	16	?	Black
3/4	19,20,24	?	Black

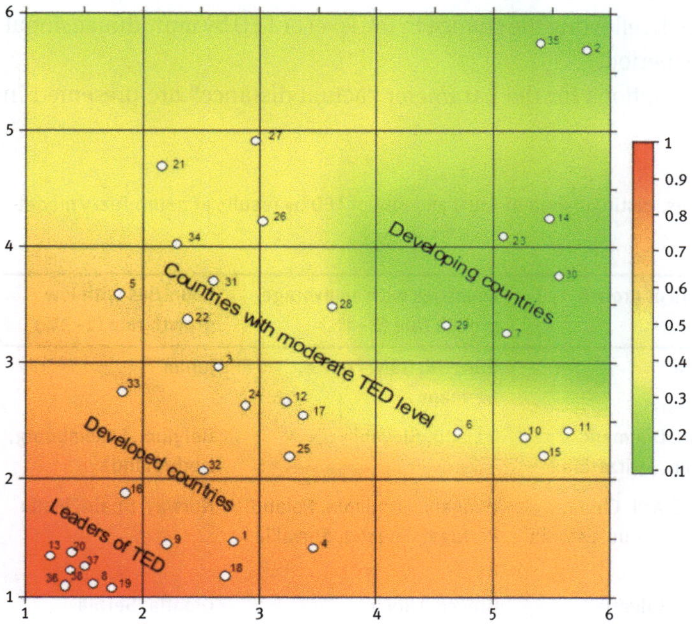

Fig. 5.16: Modified neuro-fuzzy model of world technical and economic portrait in 2010 based on FCNN-SOM

Tab. 5.5: Countries in Figure 5.16

Belgium	1	Spain	11	Lithuania	21	Russia	31
Bulgaria	2	Finland	12	Luxembourg	22	Sweden	32
Brazil	3	France	13	Latvia	23	Slovenia	33
Switzerland	4	Croatia	14	Malta	24	Slovakia	34
China	5	Hungary	15	Netherlands	25	Turkey	35
Cyprus	6	Ireland	16	Norway	26	UK	36
Greece	7	India	17	Poland	27	USA	37
Germany	8	Iceland	18	Portugal	28	Canada	38
Denmark	9	Italy	19	Romania	29		
Estonia	10	Japan	20	Serbia	30		

This set of parameters is used to calculate the level of TED and the characteristics of the individual country's fuel and energy resources and to predict future values of the multidimensional process. Their comparison on the reference and national TED trajectories makes it possible to obtain a reliable estimate not only of the speed but also of the level of the TED of each country.

Based on the results of the modified neuro-fuzzy clustering of data, the countries under study are divided into four main categories according to the level of their TED (Table 5.6), which takes a value from 1 for the reference level of TED to 0.1 for the minimum level of development. For each country, the growth rate of TED for the 5-year period is also calculated. For each 5-year period, the calculation is based on the output values of FCNN-SOM, reflecting the change in the level of TED by multidimensional characteristics for this period.

The results of calculations for the parameter "actual distance" are presented in Table 5.7.

Tab. 5.6: Country categories, distinguished by level and rate of TED by results of neuro-fuzzy modelling (2005–2010)

Group	Leaders of growth (> 4%)	Countries with an average growth rate (2–4%)	Countries with low growth rate (1–2%)
TED leaders	Canada	USA, UK, Italy, France, Germany	Japan
Developed countries	Ireland, Denmark, Sweden, Switzerland	Iceland, Slovenia	Belgium, Luxembourg, Netherlands
Countries with moderate levels of TED	India, Brazil, China, Russia, Cyprus, Estonia	Hungary, Lithuania, Poland, Portugal, Finland, Slovakia	Norway, Spain, Malta
Developing countries	Turkey, Bulgaria, Romania	Greece, Latvia	Croatia, Serbia

Tab. 5.7: Country categories identified by the parameter "actual distance" based on the results of neuro-fuzzy modeling (2010)

Actual length	Country
< 10 years	Switzerland, Finland, Germany, Denmark, Sweden, UK
10–20 years	Russia, Czech Republic, Greece, Malta, Portugal, Hungary, Lithuania, Bulgaria, Poland, Slovakia, Italy, Norway, Spain
> 20 years	Romania, Latvia, Turkey, Croatia

Conclusion

The scientific novelty of the developed architecture of the deep neural fuzzy network FCNN-SOM-RBFN consists in a successful structural and functional integration through a layer of fuzzy clustering of the advantages of cellular neural self-organization (high degree of self-organization of neurons, the possibility of selective management of individual neuronal connections for solving the problem of "dead" neurons) and multilayer RBFN. The question of the form and parameters of radially symmetric functions is mathematically justified and automatically solved in the case of normalized input data. Thus, the most difficult stage of RBFN configuration is implemented – partitioning of input attribute spaces and setting of parameters of radially symmetric functions. The concept of the new model is based on Pospelov's classification [31] based on two methods of representation of knowledge – extensional (the formation of clustered generalized images) and intensional (identification of integrated characteristics of classes in isolated clusters).

The architecture is quite compact compared to most models of pattern recognition that require very large computing power. It lacks logical relationships and aggregation of conditions in fuzzy systems.

Its practical value is in its high flexibility, ease of implementation and time efficiency of the method. High learning speed and universal approximating properties of the proposed network will be especially useful when processing multidimensional vector argument functions.

Listed in what follows are the main results obtained in this chapter.

1. The classification problem of the multidimensional overlapping of objects reduces to a fuzzy modification of pattern recognition tasks.
2. Based on the promising developments in the field of cellular self-organizing neural networks and fuzzy clustering procedures, the author proposed a deep fuzzy neural network for pattern recognition and algorithms for training, characterized by a high degree of neuronal self-organization, the absence of heuristic training parameters, the ability to control selectively individual connections between neurons to solve the problem of "dead" neurons, the high convergence rate and improvement of the separation properties of the network in the case of overlapping clusters and classes.
3. The proposed FCNN-SOM can be effectively used not only for multidimensional clustering of semi-structured data in the form of generalized patterns but also for the automatic configuration of parameters and the type of membership functions (fuzzification) in adaptive fuzzy inference systems.
4. The network can operate in the mode of exact and fuzzy clustering, depending on the nature of the data being processed.
5. To improve the accuracy of the intersecting features, a deep neuro-fuzzy classifier that combines FCNN-SOM and RBFN through a layer of fuzzy clustering is proposed.

6. An experimental study on test data on the qualitative and quantitative evaluation of the solution demonstrated the model's computational efficiency and capability for fuzzy separation of overlapping clusters.
7. The practical importance of the developed methodology of multicriteria neural network analysis of global TED was demonstrated on a set of quantitative and qualitative indicators, including technological and social dimensions of TED. Based on that, we constructed a flexible neuro-fuzzy model of the trajectory of global TED as a dynamic range of FCNN, differing in having a quite simple but powerful strategy for increasing the accuracy of clustering, without compromising interpretability, the assessment of the TED level, TED characteristics of individual countries and prediction of future values of a multidimensional process. Comparison of these parameters to leadership and national TED trajectories makes it possible to obtain a reliable estimate not only of the speed but also of the level of TED of each country.
8. To build a predictive model on a neural network is not a requirement of stationarity of the process. The neural network is a universal tool that can accurately identify non-linear patterns and relationships between the components of multivariate random processes. The use of fuzzy clustering allows a more accurate separation of overlapping clusters.

In conclusion, perspectives for future research in the field of mining multidimensional semi-structured data, describing complex intersecting objects, are associated with the development of effective models of deep fuzzy neural networks based on different principles concerning the introduction of fuzziness in the structure and on the basis of modification of teaching strategies.

Acknowledgment: This chapter was prepared within the framework of the Competitiveness Enhancement Programme of National Research, Tomsk State University, with the financial support of the Russian Foundation for Basic Research (grant 16-29-12858).

Bibliography

[1] Kohonen T. *Self-organizing maps.* Moscow, Russia: BINOM. Knowledge laboratory, 2008.
[2] Duke VA. *Computer psychodiagnostics.* Saint Petersburg, Russia: Brotherhood, 1994.
[3] Pospelov DA. *Artificial intelligence. Reference. Book 2. Models and methods.* Moscow, Russia: Radio and communication, 1990.
[4] Lutsenko EV. *Theoretical bases and technology of adaptive semantic analysis to support decision-making.* Krasnodar, Russia: CUI MVD RF, 1996.
[5] Lutsenko EV. *Automated system-cognitive analysis in the management of active objects.* Krasnodar, Russia: KubSAU, 2002.
[6] Bohr N. *Atomic physics and human knowledge.* Moscow, Russia: Mir, 1961.

[7] Gorbachev SV. Intelligent mapping system ", Informeo": the experience of using neuroinformation technologies. In *Proceedings of VIII all-Russian seminar "Neuroinformatics and its applications"*, pages 146–148. Krasnoyarsk, Russia: Publishing house Krasnoyarsk state University, 2000.

[8] Gorbachev SV. Building neural network control system to solve the problem of the identification of geological bodies 3-D. In *Proceedings of the IX all-Russian seminar "Neuroinformatics and its applications"*, pages 113–114. Krasnoyarsk, Russia: Publishing house Krasnoyarsk state University, 2001.

[9] Gorbachev SV and Syryamkin VI. *Neuro-fuzzy techniques in intelligent systems of processing and analysis of multidimensional information*. Tomsk, Russia: Publishing house Tomsk state University, 2014.

[10] Bodyanskiy Y, Gorshkov Y, Kolodyaznhiy V, and Stephan A. Combined learning algorithm for a selforginizing map with fuzzy inference. In *Proceedings of Computational intelligence, theory and applications: International Conference 8th Fuzzy Days in Dortmund*, volume 33, pages 641–650, 2005.

[11] Borisov VV, Kruglov VV, and Fedulov AS. *Fuzzy models and networks*. Moscow, Russia: Telecom, 2012.

[12] Hoeppner F, Klawonn F, Kruse R, and Runkler T. *Fuzzy Clustering Analysis: Methods for Classification, Data Analysis and Image Recognition*. Chichester: John Wiley & Sons, 1999.

[13] Keller JM and Hunt DJ. Incorporating fuzzy membership function into the perceptron algorithm. *IEEE Transactions on Pattern Analysis and Machine Intelligence*, 7:693–699, 1985.

[14] Mitra S and Pal SK. Fuzzy multi-layer perceptron, inferencing and rule generation. *IEEE Transactions on Neural Networks*, 6:51–63, 1995.

[15] Bezdek JC. *Pattern Recognition with Fuzzy Objective Function Algorithms*. New York, USA: Plenum Press, 1981.

[16] Bodyansky EV and Rudenko OG. *Artificial neural networks: architectures, learning, applications*. Kharkov, Ukraine: TELETECH, 2004.

[17] Xu R and Wunsch DC. Clustering. In *IEEE Press Series on Computational Intelligence*. Hoboken, USA: John Wiley & Sons, 2009.

[18] Gorban AN. *Training of neural networks*. Moscow, Russia: Paragraf, 1990.

[19] Vuorimaa P. Fuzzy self-organizing map. *Fuzzy Sets and Systems*, 66:223–231, 1994.

[20] Vuorimaa P. Use of the fuzzy self-orginizing map in pattern self-recognition. In *Proceedings of 3-rd IEEE Int. Conf. Fuzzy Systems "FUZZ-IEEE'94"*, pages 798–801, Orlando, USA, 1994.

[21] Tsao ECK, Bezdek JC, and Pal NR. Fuzzy Kohonen clustering networks. *Pattern Recognition*, 27:757–764, 1994.

[22] Pascual-Marqui RD. Smoothly Distributed Fuzzy cMeans: a New Self-Organizing map. *Pattern Recognition*, 34:2395–2402, 2001. R. D. Pascual-Marqui, A. D. Pascual-Montano, K. Kochi, J. M. Caraso.

[23] Anikin VI and Karmanov AA. The training of the artificial neural network Kohonen's self-organizing cellular automaton. *Information technologies,*, 11:73–80, 2014.

[24] Chua LO and Yang L. Cellular neural networks. Theory. *IEEE Transactions on Circuits and Systems CAS-35*, 10:1257–1272, 1988.

[25] Chua LO and Yang L. Cellular neural networks. Applications. *IEEE Transactions on Circuits and Systems CAS-35*, 10:1273–1290, 1988.

[26] Dvoretzky A. On stochastic approximation. In *Proceedings 3-rd Berkley Symp. Math. Statistics and Probability*, pages 39–55, 1956.

[27] Dawod M, Hasan M, and Daood A. Experimental Study: Comparison of clustering algorithms. *Int. Journal of Engineering Research and Application*, 7(8):23–34, 2017.

[28] Bezdek JC, Tsao ECK, and Pal NR. Fuzzy Kohonen Clustering Networks. *IEEE Transactions on Fuzzy Systems*, pages 1035–1043, 1992.

[29] Karayiannis NB and Pai PI. Fuzzy algorithm for learning vector quantization. *IEEE Trans. on Neural Network.*, 5:1196–1211, 1996.

[30] Wu KL and Yang MS. A fuzzy-soft learning vector quantization. *Neurocomputing is*, 55:681–697, 2003.

[31] Kondratiev ND, Yakovets YV, and Abalkin LI. *Large cycles of conjuncture and theory of foresight: Selected works*. Moscow: Economy, 2002.

[32] Glazyev S and Kharitonov V. *Nanotechnology as a key factor in the new technological order in the economy*. Moscow: Twent, 2009.

[33] Gorbachev SV and Syryamkin VI. Cognitive neural network modeling of the trajectory of global technical and economic development. In *Proceedings of International Conference on Cognitive Computing and Information Processing CCIP-2015*, pages 341–346. NOIDA, India, 2015.

Satish Nirala, Deepak Mishra*, K. Martin Sagayam, D. Narain Ponraj, X. Ajay Vasanth, Lawrence Henesey, and Chiung Ching Ho

6 Image fusion in remote sensing based on sparse sampling method and PCNN techniques

Abstract: Image fusion is a combination of multiple images that results in a fused image. It provides more information than any other input images and is based on discrete wavelet transformation (DWT) and a sparse sampling method. The sparse sampling method offers better performance than Nyquist theorem as a signal processing technique. Among the various techniques, DWT offers many advantages, such as yielding higher quality, requiring less storage and incurring lower costs, which is very useful in image applications. Image-related applications have few constraints, such as minimum data storage and lower bandwidth for communication that takes place through satellites, which actually results in the capture of low-quality images. Major technical constraints, such as minimum data storage for satellite imaging in space and lower bandwidth for communication with earth-based stations, etc., limit the ability of satellite sensors to capture images with high spatial and high spectral resolution simultaneously. To overcome this problem, image fusion has proved to be a potential tool for remote sensing applications that incorporate data from combinations of panchromatic, multispectral images and that aim to produce a composite image having both higher spatial and spectral resolutions. Research in this area can be traced back to the most recent couple of decades; unfortunately, the diverse methodological approaches proposed so far by different researchers have not been discussed. Experimental results indicate that a comparative analysis of two methods used in image fusion, such as compressive sensing and pulse-coupled neural network (PCNN), for a few remote sensing **location data**, e.g., England, forest land, Egypt, island, Balboa and bare soil, is used in this work. The following performance metrics have two assessment procedures, i.e., at full- and reduced-scale resolutions, to evaluate the performance of these algorithms.

Keywords: Compressive sensing, sparse representation, discrete wavelet transform, PCNN

Satish Nirala, Deepak Mishra, Department of Avionics, Indian Institute of Space and Technology, Trivandrum, India
K. Martin Sagayam, D. Narain Ponraj, X. Ajay Vasanth, School of Engineering and Technology, Karunya University, Coimbatore, India
Lawrence Henesey, Department of Systems and Software Engineering, Blekine Institute of Technology, Sweden
Chiung Ching Ho, Department of Computing and Informatics, Multimedia University, Malaysia

https://doi.org/10.1515/9783110551433-006

6.1 Introduction

Nowadays, image fusion has become a very interesting research topic involving many related areas such as automatic object detection, computer vision, medical imaging, remote sensing, robotics and night vision applications. Furthermore, the need for image fusion in computer vision systems is increasing dramatically. Image data must be extracted and combined from multiple images from the same scene. The result of image fusion is a single image, which is more appropriate and feasible for both humans and machines to observe for further image processing tasks. Multifocus and multisensor image fusion are the two important branches of image fusion. Due to the limited depth of field of optical lenses, it is difficult to obtain an image with all objects in focus. Multifocus image fusion attempts to increase the apparent depth of field through the fusion of objects within several different fields of focus. In many situations, a single sensor is unable to provide an exact view of real-time applications. Thus, multisensor data fusion has emerged as an extraordinary research area in recent years. Multisensor fusion is a process of binding data from various sensors into a solitary representational format. Image fusion methods are divided into two major classes: spatial domain image fusion and transform domain image fusion. Spatial domain image fusion includes fusion methods like principal component analysis (PCA), which is based on a fusion method, averaging-based image fusion, and minimum and maximum information-based image fusion methods. Transform domain image fusion includes fusion methods based on multiscale transform such as dual-tree complex wavelet transform (DTCWT) and discrete wavelet transform (DWT). PCNN was developed by Eckhorn in 1990 and based on experimental observations of synchronous pulse bursts in cat and monkey visual cortex. It is characterized by the global coupling and pulse synchronization of neurons. These features facilitate image fusion, which makes use of local image information. All fusion the aforementioned methods operate on the basis of whole sampling, which requires very high computational and high-complexity computer resources. It is a difficult problem for the increasing real-time demand placed on image fusion systems.Then compressed sensing (CS) is used to provide a way to resolve the problem of image fusion. CS theory enables a sparse or compressible signal to be properly reconstructed from a small number of non-adaptive linear projections. For image fusion at the pixel level, a small amount of measurements can be directly fused in the compressed domain and be reconstructed with high accuracy according to the fused measurements. It will greatly reduce the demand for computational resources and complexity. The implemented image fusion methods include the PCA method and the spatial or simple averaging method in the spatial domain category. DWT- , DTCWT-) and PCNN-based image fusion methods were applied in the transform domain category. Image fusion in a CS framework was also implemented, which includes the aforementioned transformed domain image fusion methods. Integrated image fusion, a technique based on PCNN and PCA with or without a CS framework, is proposed in this chapter. The proposed method gives clearer results methods currently used in the literature since the fusion process is carried out based on the different properties of high-pass- and low-pass-frequency components of images.

We focus on the results of an investigation into an objective evaluation of an image fusion method. The gradient-based objective fusion framework was tested by C. Xydeas and V. Petrovic [1], who make inferences on input gradients of the fused images under evaluation. The results and discussion section presents an analysis of image fusion methods evaluated using quality measures such as entropy (E), standard deviation (SD), mutual information (MI) and others.

6.1.1 Motivation and scope

In the field of image processing a lot of developments have taken place recently, where image fusion is playing a vital role in research on night vision, military applications and others. Many imaging devices suffer from a limited depth of field, so there is a need for the development of better-quality fused images in multifocus image fusion, as suggested by S. Wei and W. Ke [2]. The works related to image fusion algorithms indicate that all existing algorithms degrade the brightness of fused images and are time consuming. Since image features differ from each other, we need a fusion method based on multiscale decomposition so that all the information in decomposed images can be acquired and fused.

In this chapter, an integrated image fusion method is developed and implemented. Image fusion using PCNN and PCA in a compressive sensing approach, where PCNN and PCA methods are used for fusing low-pass frequency and high-pass frequency component respectively. The performance measures from an experimental result improves of evaluation and visual effect metrics. It has been compared with the other conventional methods based on the frequency components of the image. The fusion results obtained using the proposed methods show a significant improvement in image quality visually and in terms of the evaluation metrics, since the fusion process is done based on the different properties on high pass and low pass frequency components of image.

6.1.2 Literature survey

To date, many image fusion methods have been developed that vary in complexity and applications. The main objective of some of the popular image fusion techniques will be discussed in this section.

C. Polh and J. Van Genderen [3] discussed image fusion using PCA (PCA). PCA is used to extract the weighted feature points in image fusion. The principal components of the images are the eigenvectors corresponding to the largest eigenvalues of the co-variance matrix formed by the images. Hence, it reduces redundancy in the fused image data.

H. Yin et al. [4] presented an image fusion technique based on a sparse sampling method or CS. The execution rate is reduced by increasing the quality of the resultant image fused by the integration of non-zero coefficients present in the image. Many errors occur in the conventional approach, such as reconstruction errors, poor fidelity and blocking artefacts that produce uncertain results. To overcome these problems, another fusion-based approach was proposed that includes non-subsampled contourlet transform, which is widely used in decomposition techniques. The CS-based fusion approach can greatly reduce processing speed and guarantee the quality of fused images by integrating fewer non-zero coefficients. Due to the limitations in the conventional CS-based fusion approach, directly fusing sensing measurements may produce results that have greater uncertainty with a high rate of reconstruction errors, and using a single fusion rule may result in problems of blocking artefacts and poor fidelity. An image fusion approach based on CS was developed to solve these problems. The NSCT has been used for image decomposition. A dual-layer PCNN model was used to carry out low-pass sub-band fusion, while high-pass sub-band fusion was used to implement the edge-based fusion rule. For the reconstruction of fused images, a CS matching pursuit algorithm (CoSaMP) was propsed by D. Needell and J. A. Tropp [5].

The motivation behind using PCNN, used to transfer the spatial frequency domain using NSCT for an image fusion procedure, is discussed in Q. Xiao-Bo [6]. Non-subsampled contourlet transform (NSCT) provides flexible multiresolution, anisotropic and directional expansion of images. Compared with the original contourlet transform, it is shift-invariant and can overcome the pseudo-Gibbs phenomenon around singularities. The PCNN is a visual cortex-inspired neural network characterized by the global coupling pulse synchronization of neurons. It has proven suitable for image processing and successfully used in image fusion. Q. Xiao-Bo et al. [6] have shown through experimentation that **NSCT** is associated with PCNN and can be used in image fusion to make full use of image characteristics. Spatial frequency in the NSCT domain is used to motivate PCNN, and coefficients in the NSCT domain with large firing times are selected as coefficients of fused images. The results obtained from experiments show that the algorithm has performed well using contourlet-based PCNN and wavelet-based PCNN in terms of visual quality and performance metrics.

Yunxiang Tian and Xiaolin Tian [7] have proposed to extend the work of [6]. This includes the intensity hue saturation (IHS) space component from multispectral (MS) images, then decomposition into an I component using NSCT. This method reduces computation time and complexity. Yang Chen and Zheng Qin [8] have presented PCNN-based image fusion in a compression domain approach. In this experiment, the researchers validated the performance measure for three different fusion applications. This produced results in both qualitative and quantitative analysis carried out by Vaibhav R. Pandit and R. J. Bhiwani [9] for applications in image fusion. It has been compared with various conventional image fusion techniques. Hongguang Li et al. [10] have proposed a novel method for image registration and image fusion that has been integrated with cameras in both visible and infrared light on unmanned aerial vehicles (UAVs). This system is very efficient and uses a particle swarm optimization (PSO) hybrid with PCNN and NSCT. Thus, it can be applied to medium- or high-altitude UAVs. Yong Yang et al. [11] have presented two major issues related to multifocus image fusion using fast discrete curvelet transform (FDCT) texture selection and block effects. Experimental results show a reduction in computational complexity and improvement in performance metrics, which has been compared with various state-of-the-art frequency-based fusion and spatial-based fusion methods. B. Rajalingam and R. Priya [12] have implemented, using PCA-DWT, a technique to improve medical image fusion. It is superior to other existing techniques in terms of image quality, visual perception and computational time. Biswas et al. [13] proposed an idea for remote sensing image fusion based on shearlet transform (ST) and PCNN. It can be used to vary the decomposition level of source information in order to maintain geometrical and spectral information. Jun Li et al. [14] discussed image fusion methods applied to infrared and visible images using robust PCA (RPCA) and CS. It preserves background information from infrared and visible images. It computes in parallel in less computation time.

6.2 Methodology used

The integration of various data that could provide valuable information is the major objective of image fusion. The major objective of image fusion is to integrate different data in order to obtain more information than can be derived from each of the single data. The integration of images from different scenarios will lead to additional information that should not be present in a separate imaging scenario. But owing to design or observational constraints, imaging devices are not capable of providing such information. For this reason several image fusion methods have been developed. Individual images are integrated to obtain a single image, so that all information can be obtained. The objective of the method is to design a system that satisfies the following criteria:

1. Extract all useful information from the source information.
2. Avoid artificial facts or inconsistencies that perturb human observers.
3. Be reliable and robust to imperfections such as misregistration.
4. Suppress noise and irrelevant paths.
5. Reduce inconsistencies or distortions.
6. Limit bandwidth and storage capacity.

6.2.1 Principal component analysis theory

The aim of PCA is to convert a set of observations of correlated variables into a set of values that are linearly uncorrelated variables in a systematic process called principal components. Consider X represents an m-dimensional random vector, where V represents an orthonormal projection matrix such that $Y = V^{\mathrm{T}}X$. The covariance of Y is $\mathrm{cov}(Y)$, which is a diagonal matrix, and $V^{-1} = V^{\mathrm{T}}$:

$$\begin{aligned}\mathrm{cov}(Y) &= E(YY^{\mathrm{T}})\\ &= E[(V^{\mathrm{T}}X)(V^{\mathrm{T}}X^{\mathrm{T}})]\\ &= E[(V^{\mathrm{T}}X)(X^{\mathrm{T}}V)]\\ &= V^{\mathrm{T}}E(XX^{\mathrm{T}})V\end{aligned}$$

$$\mathrm{cov}(Y) = V^{\mathrm{T}}\,\mathrm{cov}(X)V\,. \tag{6.1}$$

Multiplying both sides of Equation (6.1) by V we get

$$V\,\mathrm{cov}(Y) = VV^{\mathrm{T}}\,\mathrm{cov}(X)V = \mathrm{cov}(X)V\,. \tag{6.2}$$

V can be written $V = [V_1, V_2, \ldots V_d]$ and

$$\begin{bmatrix} \lambda_1 & 0 & \ldots & 0 \\ \ldots & \ldots & \ldots & \ldots \\ 0 & 0 & \ldots & \lambda_d \end{bmatrix} \tag{6.3}$$

Substituting Equation (6.2) into Equation (6.3)

$$[\lambda_1 V_1, \lambda_2 V_2 \ldots \lambda_d V_d] = [\mathrm{cov}(X)V_1, \mathrm{cov}(X)V_2 \ldots \mathrm{cov}(X)V_d]\,, \tag{6.4}$$

which can be written

$$\lambda_i V_i = \mathrm{cov}(X)V_i, \tag{6.5}$$

where $i = 1, 2\ldots d$, and V_i is an eigenvector of $\mathrm{cov}(X)$.

6.2.1.1 PCA image fusion algorithm

Step 1: First we will generate a two-column matrix A of dimension $2 \times N$ using the two source images and calculate the empirical mean vector along each column M_{e} of dimension 1×2.

Step 2: Subtract M_{e} from each column of A; the resulting matrix X will be of dimension $2 \times N$.

Step 3: Find the covariance matrix C of X and calculate the eigenvectors V and eigenvalue D of C and arrange them in decreasing order of their eigenvalues.

Step 4: Calculate the normalized principal components P_1 and P_2 as follows:

$$P_1 = \frac{V_1}{\Sigma V} \quad \text{and} \quad P_2 = \frac{V_2}{\Sigma V} . \tag{6.6}$$

Step 5: The fused image is

$$I_{\text{fuse}}(x, y) = P_1 I_1(x, y) + P_2 I_2(x, y) . \tag{6.7}$$

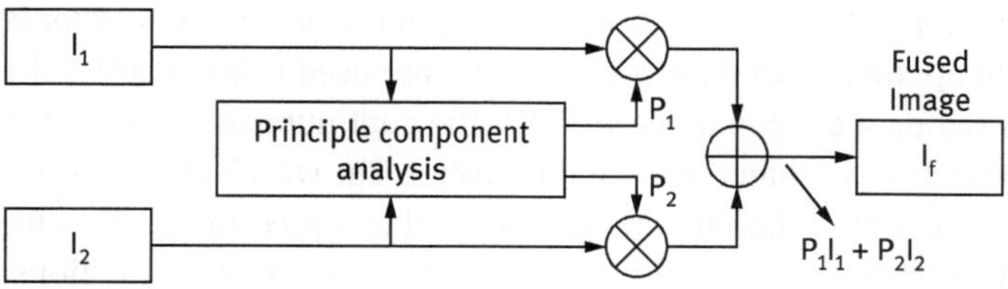

Fig. 6.1: Information flow diagram in image fusion scheme employing PCA [12]

6.2.2 Spatial or simple averaging

The average value of the input image is taken pixel by pixel using a simple image fusion method. The averaging-based fusion rule is as follows:

$$I_F(i, j) = \frac{I_A(i, j)}{2} + \frac{I_B(i, j)}{2} , \tag{6.8}$$

where I_F is the fused image and I_A, I_B are source components of the image. Each image has three basic multispectral bands of R, G and B.

The relative averaging of two groups, C_X and C_Y, which produced C_F, has a single band for all individual pixels, represented in Equation (6.9):

$$C_F(m, n, i) = \frac{C_X(m, n, i)}{2} + \frac{C_Y(m, n, i)}{2} , \tag{6.9}$$

where (m, n) the denotes pixel and spatial domain and i denotes the R, G, B of an image.

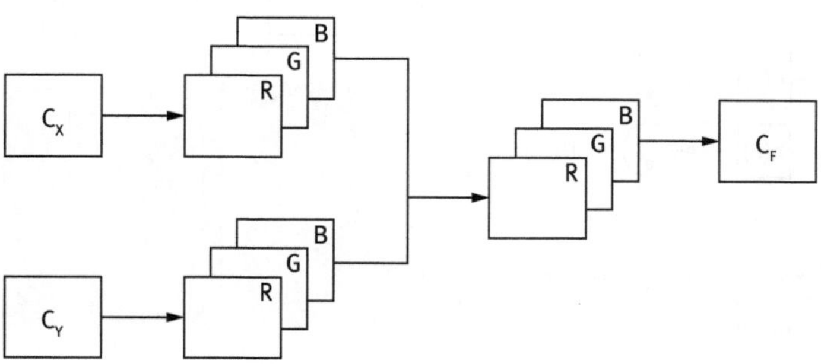

Fig. 6.2: Block diagram of spatial averaging method [12]

6.2.3 Image fusion–based wavelet transform

6.2.3.1 Discrete wavelet transform

Signals can be represented with higher efficiency using wavelet transforms, which is one of the popular existing techniques. A signal is transformed into a set of basis functions, referred to as a wavelet. The wavelet is chosen and the level of decomposition is chosen at the initial stage. A threshold value is determined after the decomposition of the input data. After applying DWT, the size of the resultant output data vector is the same as that of the applied input data vector. Filters are used to reconstruct the original data. The filtered data are compared to DWT, the continuously transformed data are returned that were larger than the input data vector. The wavelet transform is a well-known signal processing technique that can be used to represent signals with high efficiency. The wavelet transform changes signals into a set of basis functions. These basis functions are called wavelets. A waveform of finite duration having an average value of zero is called a wavelet. A multiscale representation of a signal is given by a DWT. Most applications are not directly associated with scaling factors or wavelet coefficients. A filter band coefficient filters input signals using high-pass and low-pass filters. Let the low pass be $h(n)$ and high pass be $g(n)$. Many sets of filters satisfy the perfect reconstruction conditions, which are approximately symmetric. To use a two-dimensional transform image for analysis, synthesis filter bands can be implemented. In the 2D case, a 1D analysis filter band is initially connected to a segment of the image and then connected to the rows. Let the image representation have N_1 rows and N_2 columns; then, after applying the 1D analysis filter band to each column, we have two subband images, each with $N_1/2$ rows and N_2 columns; after applying the 1D analysis filter band to each row of both subband images, we have four subband images, each having $N_1/2$ rows and $N_2/2$ columns. This is illustrated in Figure 6.3, which shows the output of the DWT of the image and its analysis block.

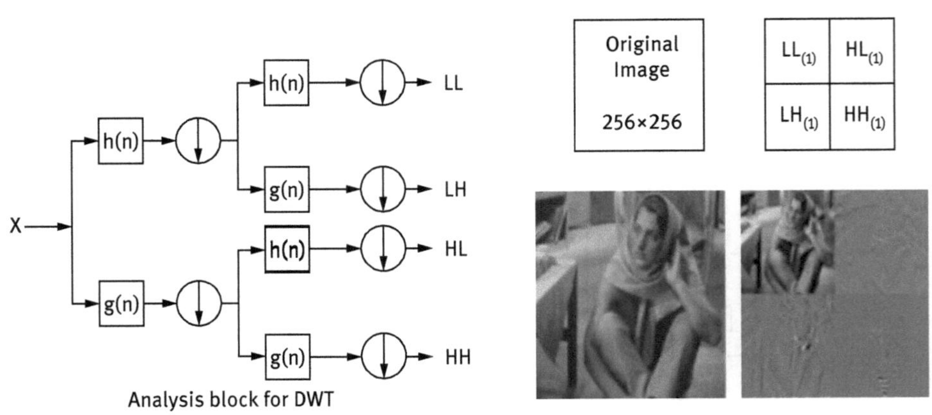

Fig. 6.3: DWT and its output image

6.2.3.2 Dual-tree complex wavelet transform

Selesnick et al. [15] have shown how DTCWT is a more advanced form of wavelet transform compared to DWT. There are two major advantages of DTCWT over DWT, shift invariant and selective directional into higher dimensions. The major shortcomings in DWT are as follows.

1. Wavelet coefficients tend to oscillate between positive and negative around singularities because of the wavelet bandpass function.
2. A small shift in a signal perturbs the wavelet coefficient oscillation pattern around singularities.
3. The lack of directional selectivity greatly complicates the modelling and processing of geometric image features.

One of the advantages of the DTCWT is that it can be used to implement 2D wavelet transforms, which are more selective with respect to the orientations than in separable 2D DWTs. There are two types of 2D DTCWT: the real 2D DTCWT tree is twice as expensive, while the complex 2D dual-tree DWT is four times as expensive. Both types of complex wavelet transforms have wavelets that are oriented in six distinct directions. Six wavelets are spread in distinct directions. Each subband of the 2D dual-tree transform indicates a specific orientation. A complex 2D dual tree is implemented as four critically sampled and separable 2D DWTs operating in parallel. However, different filter sets are used in the rows and columns. As in real cases, the sum and difference of subband images are determined to obtain the oriented wavelets. A real 2D dual-tree transform of an image is implemented using two critically sampled and separable DWTs operating in parallel. The sum and difference for each subband are taken into consideration. As shown in Figure 6.4, a complex 2D DTCWT is implemented as four critically sampled and separable 2D DWTs operating in parallel. The pair of conjugate filters applied to image decomposition can be expressed as

$$(H_x + jG_x)(H_y + jG_y) = (H_xH_y - G_xG_y) + j(H_xG_y + G_xH_y)\,. \tag{6.10}$$

The complex conjugate of the wavelet transform has positive and negative coefficients of subbands. All corner-to-corner subbands are partitioned in both vertical and horizontal scale of six subbands in the block.

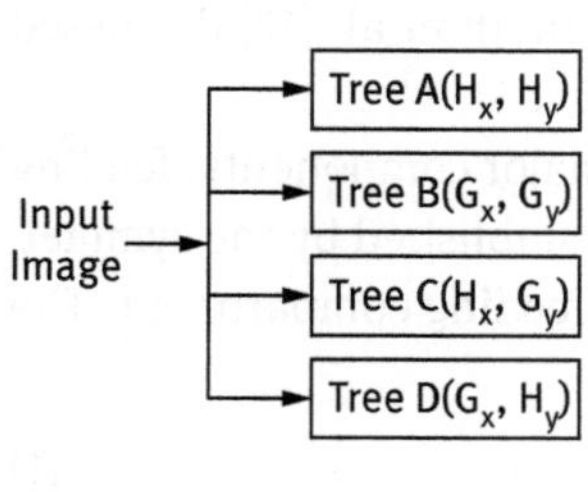

Fig. 6.4: Analysis block for DTCWT

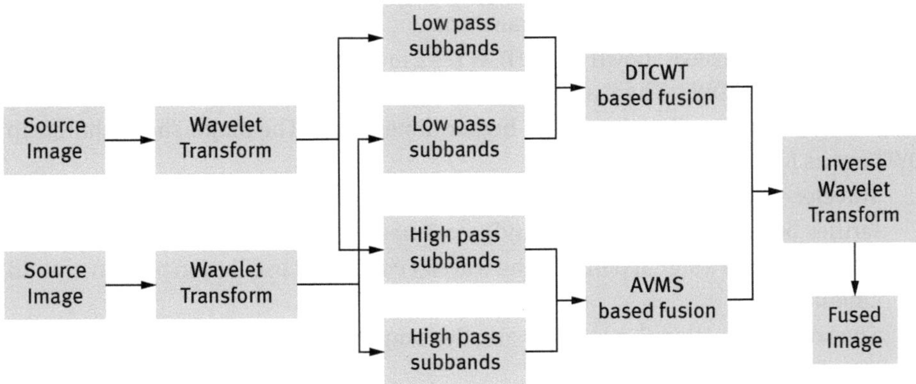

Fig. 6.5: Block diagram of wavelet transform–based image fusion method

Figure 6.5 shows a general block diagram of a wavelet transform–based image fusion method. It consists of low-pass and high-pass subbands for the fusion process in terms of averaging-based and average mean square–based fusion techniques. The fused images must be obtained by taking the inverse wavelet transform from the previous step.

6.2.4 Pulse-coupled neural network

This works on the principle of optic nerves, which transmit data observed by the eyes to the brain and the visual cortex. Generally, the input data are then transferred to fiber pathways and finally processed. The most predominant factor of the visual cortical area in the visual cortex is visual processing, and nearly all visual signals reach the cortex via the primary visual cortex. Eckhorn et al. [16] in the late 1980s discovered that binary images are created in an oscillating manner by the mid-brain, which can extract different features from the visual impression when they have to read the cat visual cortex [16] An actual image is then created in the cat's brain by the help of these binary images. Inspired by this biological system, machines are also trained to perform the functions of the human brain. This leads to the new structure of a neural network model, called Eckhorn's model, to simulate the behaviour of the visual cortex. From the case study report on the linking field model H. Ranganath et al. [17] discussed PCNN based on the Eckhorn model with some modifications.

Figure 6.6 shows a PCNN model that consists of two major components: feeding and linking. The link between neighbourhood neurons is established by the symmetric weights M and W. The input stimulus S is given to the feeding compartment. The PCNN model and its components are determined by 6.11

$$F_{ij}[n] = e^{-\alpha F} F_{ij}[n-1] + S_{ij} + V_F(M * Y)\,, \tag{6.11}$$

$$L_{ij}[n] = e^{\alpha L} L_{ij}[n-1] + V_L(W * Y)\,, \tag{6.12}$$

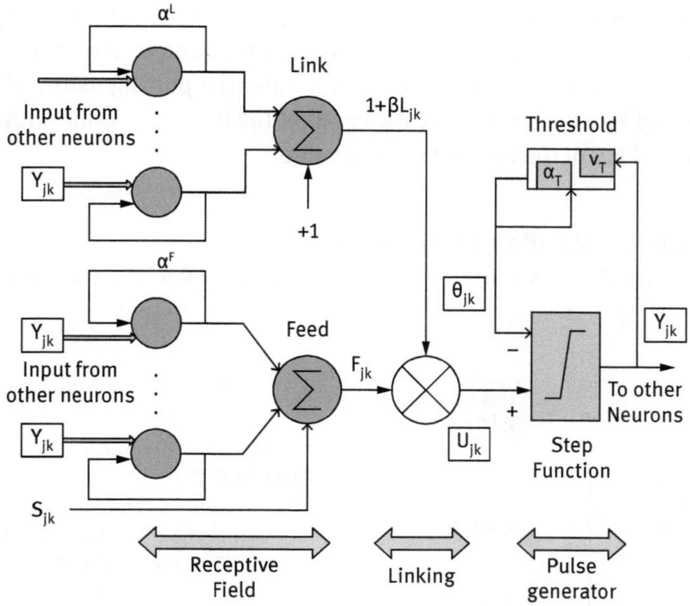

Fig. 6.6: Schematic representation of PCNN processing element [18]

where $*$ represents the convolution operator, 'n' represents the number of iterations, 'Y' represents the output of neurons from $(n-1)$ iterations. The compartments have a memory of their previous state, which decays over time by the exponent term. The constants V_F and V_L are normalizing constants. After receiving the inputs the feeding and linking compartments are combined together to generate the internal state of the neurons. The internal activity U is expressed as in Equation (6.13), where β is the controlling parameter:

$$U_{ij}[n] = F_{ij}[n](1 + \beta L_{ij}[n]). \tag{6.13}$$

Thus, the internal activity of neuron U has various states mapped with the dynamic threshold, θ, to produce output to another neuron 'Y' as given in Equations (6.14) and (6.15):

$$Y_{ij}[n] = 1\,; \quad \text{if } U_{ij}[n] > \theta_{ij}[n]\,, \tag{6.14}$$

$$Y_{ij}[n] = 0\,; \quad \text{Otherwise}\,. \tag{6.15}$$

If $(U > \theta)$, then the neuron fires very high to the threshold value of Y and converges until a neuron gets fired. The dynamic activity of the threshold property is given in Equation (6.16),

$$\theta_{ij}[n] = e^{\alpha\theta}\theta_{ij}[n-1] + V_\theta Y_{ij}[n]\,, \tag{6.16}$$

where V_θ is a constant with a significantly larger value than the average value U and order of magnitude. Initially, values of F, L, U and Y are all set to zero. The values of the

θ elements are initially 0 or some larger value depending upon the application. From this activity, in both the feeding and linking compartments, the neuron receives some source stimulus S and stimulus from the neighbourhood state. The pulsing nature of the PCNN can be observed by looking at the rise and fall of the threshold value with the corresponding rise and fall of the internal activity value.

6.2.4.1 General image fusion algorithm based on PCNN

A data flow diagram of the PCNN algorithm is shown in Figure 6.7 and follow the steps given below,

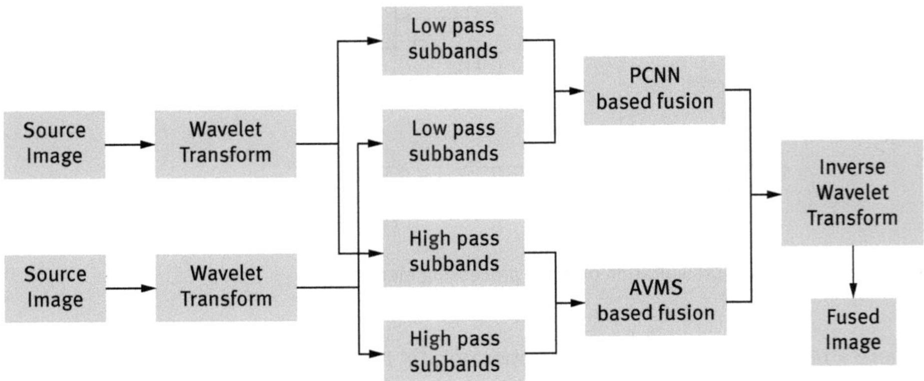

Fig. 6.7: Data flow diagram of PCNN based image fusion

Step 1: Segment input data into subbands of low pass and high pass and are called source data.

Step 2: The focus measure for the low-pass subband is given as an input to the PCNN to activate the neural activity using spatial focusing.

Step 3: Calculate the firing times T_{ij} using the following equations:

$$F_{ij}[n] = S_{ij}\,, \tag{6.17}$$

$$L_{ij}[n] = e^{\alpha l}L_{ij}[n-1] + V_L(WXY)\,, \tag{6.18}$$

$$U_{ij}[n] = F_{ij}[n](1 + \beta L_{ij}[n])\,, \tag{6.19}$$

$$\theta_{ij}[n] = e^{\alpha}_{ij}[n-1] + VY_{ij}[n]\,, \tag{6.20}$$

$$Y_{ij}[n] = 1\,; \quad \text{if} \quad U_{ij}[n] > \theta_{ij}[n]\,, \tag{6.21}$$

$$Y_{ij}[n] = 0\,; \quad \text{Otherwise}\,, \tag{6.22}$$

$$T_{ij} = T_{ij}[n-1] + Y_{ij}[n]\,. \tag{6.23}$$

We can take the parameters as

$$\alpha L = 0.06931\,, \quad \alpha\theta = 0.2\,, \quad \beta = 0.2\,,$$
$$V_L = 1\,, \quad V = 20\,, \quad W = \begin{bmatrix} 0.707 & 1 & 0.707 \\ 1 & 0 & 1 \\ 0.707 & 0 & 0.707 \end{bmatrix}$$

and the iteration is $n = 200$.

Step 4: Obtain the decision map $D_{F,ij}$ and select the coefficients $x_{F,ij}$ which means that the coefficients with large firing times (large T values) are selected as coefficients of the fused image, since the value of each pixel in the firing map images is equal to the firing times, and a longer firing time corresponds to more information, based on the following equations:

$$D_{F,ij}[n] = 1\,; \quad \text{if} \quad T_{1,ij}[n] \geq T_{2,ij}[n]\,, \tag{6.24}$$
$$D_{F,ij}[n] = 0\,; \quad \text{if} \quad T_{1,ij}[n] < T_{2,ij}[n]\,, \tag{6.25}$$
$$X_{F,ij} = x_{1,ij}[n]\,; \quad D_{ij}[n] = 1\,, \tag{6.26}$$
$$X_{F,ij} = x_{2,ij}[n]\,; \quad D_{ij}[n] = 0\,, \tag{6.27}$$

where $X_{F,ij}$, $x_{1,ij}$ and $x_{2,ij}$ denote the low-pass subband coefficients of the fused image and two source images respectively.

Step 5: High-pass subband coefficients are fused using the maximum variance rule, i.e., select the coefficient of the subband that has the maximum variance or absolute value maximum selection (AVMS) fusion rule or Laplacian mixture model (LMM) – based fusion rule has been reported by Unni V. S. [19].

Step 6: Fused images are reconstructed using the selected fused coefficients via inverse wavelet transform.

6.2.5 Compressive sensing

The Nyquist-Shannon sampling theorem states that the reconstructed signal rate is higher than twice the maximum frequency. The sampling rate of some signals becomes very high due to their high bandwidth, for example, due to the acquisition of images. Therefore, the need for CS comes into the picture, where signals are compressed and represented in a small number of coefficients. Islam et al. [20] discussed the acquisition of information using CS without sampling input signals. If a signal is a "sparse sample", thenthe acquired reconstructed signal can be accurate for all calculations. The sparsity of the signal is defined on a similar basis, such as DCT or wavelet transform. Incoherence between signals and measurements plays an important role where the information developed is widely spread out. CS provides an accurate measurement for reconstructed compressive sensed output in a non-adaptive manner [19]. In CS first we generate a sparse representation of the signal. Then we generate some measurements of dimension M, using a measurement matrix of dimension $M \times N$. Finally, the reconstruction is carried out from these measurement.

Suppose we have $f = \{f_1, f_2 \dots f_N\}$ of N real-value samples of a signal, which can represented by the transform coefficients x. That is:

$$f = \psi x = \sum_{i=1}^{N} x_i \psi_i , \qquad (6.28)$$

where $\psi = \{\psi_1, \psi_2 \dots \psi_N\}$ is an $N \times N$ transform basis matrix, which determines the domain where the signal is sparse, and $x = \{x_1, x_2 \dots x_N\}$ is an N dimensional coefficient vector with $x_i = \langle x, \psi_i \rangle$. We assume that x is K-sparse, meaning that there are only significant elements in x with $K \ll N$.

Suppose a general linear measurement process computes inner products $M < N$ between f and a collection of vectors ϕ_j giving $y_j = f, \phi_j > i = 1 \dots N; j = 1 \dots M$.

If ϕ denotes an $M \times N$ matrix with ϕ_j as row vectors, then the measurements $y = (y_1, y_2 \dots y_n)$ are given by

$$y = \phi f = \phi \psi x = Ax , \qquad (6.29)$$

where y is an M-dimensional observation vector and ϕ is an $M \times N$ random measurement matrix and $A = \phi\psi$ is called the measurement matrix. We can see that CS reduces the signal of N dimension to an observation signal of M dimension, so a dimension reduction takes place for the accurate reconstruction of the signal, and the sensing matrix ϕ should satisfy the restricted isometric property (RIP) explained by R. G. Baraniuk et al. [21, 22]. In other words, the K-RIP ensures that all submatrices of ϕ of size $M \times K$ are close to isometry, and therefore the distance (and information) preserving. Random matrices are an example of measurement matrix whose entries are **i.i.d (identically distributed)** Bernoulli and Gaussian matrices. The signal reconstruction problem uses y to reconstruct the N-length signal, x that is K-sparse, given ϕ and ψ since $M < N$, it is an ill condition problem and there are infinite x' satisfying $Ax' = y$. To solve ill-conditioned problems of this kind, a conventional approach is used to minimize the l_p norm. Define the l_p norm vector x as $lx_p = \sum_{i=1}^{N} x_i p$ for $1 \leq p < \infty$ and $lx_p = \max |x_i|$. For example, consider the l_2 norm to solve this ill condition. The optimization problem is given by

$$x = \operatorname{argmin} lx_2 \quad \text{subject to} \quad Ax' = y . \qquad (6.30)$$

Many algorithms have been generated to solve these CS problems. They include linear programming (LP) techniques, greedy algorithms, gradient-based algorithms and iterated shrinkage algorithms. Although l_1 and l_2 norm has strong possibilities for exact recovery, it has disadvantages in terms of computational cost and implementation complexity. Orthogonal matching pursuit (OMP) algorithms are developed to address these problems.

6.2.5.1 Block compression sensing

In block CS, random matrices are derived from the sampling of an image block by block, suggested by M. Y. Baig et al. [18]. Consider an $I_c \times I_r$ image with $N = I_c I_r$ pixels in total. The image is split into n small blocks with dimension $B \times B$ each and sampled with the same matrix in block CS. Let x_i represent the vectorized signal of the i-th block through raster scanning and $i = 1 \dots n$, $n = N/B^2$. Construct a $B^2 \times B^2$ **i.i.d. (identically distributed)** Gaussian matrix, and by obtaining orthogonal matrix Θ by orthonormalizing the Gaussian matrix, $M_B (\ll B^2)$ rows are picked randomly from Θ to produce $M_B \times B^2$ measurement matrix ϕ_B. The corresponding y_i is expressed in Equation (6.31):

$$y_j = \phi_B x_i\,; \quad i = 1, 2 \dots n\,. \tag{6.31}$$

To get $M(n \times M_B)$, compress the sensed output y of an entire data. The output of the previous stage is given as input block by block in cognition through the matrix ϕ, given in Equation (6.32):

$$\phi = \begin{bmatrix} \phi_b & & \\ & \phi_b & \\ & & \phi_b \end{bmatrix} \tag{6.32}$$

Block CS is memory efficient, since we just need to store an $M_B \times B^2$ Gaussian matrix, rather than a full $M \times N$ one, and block CS also has several merits. The first merit is that the encoder does not consume more time for computation since linear projection sends each block for all pixels from the image. Second, the calculation is feasible due to the smaller value of ϕ_B by an initial approximation method.

6.2.5.2 SPL reconstruction algorithm

The reconstruction process is done using smooth projected Landweber (SPL) iteration. The wiener filtering was constructed using a Landweber approach for the purpose of removing blocking artefacts and noise. The approximation to the image at iteration $K + 1$, $x(k + 1)$, is produced from $x(k)$ as

$$i^{(k)} = i^{(k)} + \Phi_B^{\mathrm{T}}(yi^{(k)} - \Phi_B i^{(k)})\,, \tag{6.33}$$

$$x^{(k)} = \Psi_i(k)\,, \tag{6.34}$$

$$x^{(k)} = \mathrm{Threshold}(x^{(k)}, \lambda)\,, \tag{6.35}$$

$$x^{(k)} = \Psi^{-1} x^{(k)}\,, \tag{6.36}$$

for each block i,

$$x^{(k+1)} = x^{(k)} + \Psi_B^{\mathrm{T}}(yi^{(k)} - \Phi_B i^{(k)})\,, \tag{6.37}$$

we terminate the iteration when $|D^{k+1} - D^k| < \epsilon$, where

$$D^k = \frac{1}{\sqrt{N}} \|\overline{x^{(k)}} - \overline{x^{(k-1)}}\|_2\,. \tag{6.38}$$

6.2.5.3 Thresholding

Projected Landweber (PL) algorithm uses hard thresholding for denoising. For the purpose of setting a proper Γ for hard thresholding, the universal threshold method is used:

$$x^{(k+1)} = x^{(k)}\,; \quad |x^{(k)}| \geq \lambda\sigma^{(k)} \tag{6.39}$$

$$= 0\,; \quad \text{otherwise}\,. \tag{6.40}$$

The parameter $\sigma^{(k)}$ is estimated using a robust median estimator,

$$\sigma^{(k)} = \frac{\text{median}(|x|)}{0.6745}\,. \tag{6.41}$$

6.3 An integrated approach to image fusion

The main objective of fusion techniques is to identify the most essential elements of an image and to retain them in the original image. This transformed into smaller subbands without any loss of information. The demands of an effective image fusion technique lies in the quality of the image, computational complexity and computational time. Based on different applications, there are varying demands on computation time, complexity, environmental constraints and information quality. The main requirement of a fusion method is that it must be able to identify essentially the largest elements in input images and transmit them without missing any elements in the fusion image. The disadvantage of the spatial domain approach is the production of spatial distortion in the image fusion. The required results obtained with spectral distortion are worst, while predicting the image from the database. The problem of spatial distortion can be overcome by transforming domain approaches to image fusion. The image details and directional coefficients are taken into account when multiscale transform-based fusion is performed. The relative coefficients of an image are fused by different sets of rules for relevant coefficients. The conventional approach over-

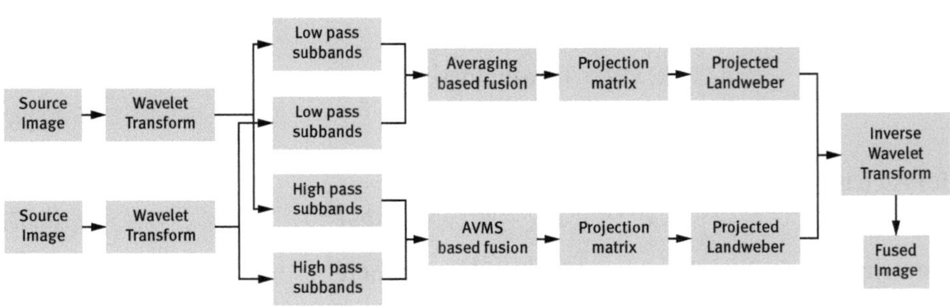

Fig. 6.8: Block diagram of CS-based image fusion method [20]

comes the cost values and the problem of higher computational complexity. A multi-scale transform-based fusion approach takes image details and direction coefficients into account. It can accomplish high fused image quality. According to the properties of different coefficients, different effective fusion rules are required to fuse the relevant coefficients. For the purpose of avoiding low computational complexity and storage or cost reduction problems, a conventional CS-based fusion approach is required. Based on the aforementioned facts an integrated image fusion method is proposed that separately deals with the low-pass- and high-pass-frequency subbands and reduces the storage and computational problem via a CS framework.

6.3.1 Image fusion using PCNN and PCA

In the proposed method, images are initially decomposed into low-pass subbands and high-pass subbands using wavelet transform. DWTs are used for image decomposition. The high-frequency subbands contain the coefficients of an image, which have both real and imaginary terms corresponding to sharp depth changes and salient features in the input data. The energy of the high-frequency coefficients of a clear image is much larger than that of a blurred one. Therefore, the variance of the focused image is greater than that of a blurred image. We will be using this variance property for fusing high-frequency subbands. We will be using a PCA approach for the fusion process. For low-pass-frequency subband fusion we will use a PCNN model-based fusion approach as it incorporates the human visual perception of images and utilizes neighbourhood information. This PCNN-based image fusion is done based on activity-level measurement.

6.3.2 Algorithm

Step 1: Consider the input images as I_A and I_B.

Step 2: Decompose the input images into high-pass and low-pass subbands as I_{LA}, I_{LB} and I_{HA}, I_{HB} respectively using wavelet transform. For example, use DWT.

Step 3: Fuse the low-pass subband using a PCNN model-based fusion rule as mentioned earlier and obtain a fused low-pass subband image as I_{LF}.

Step 4: Fuse the high-pass subbands using a PCA-based fused algorithm as mentioned earlier, and obtain a fused high-pass subband as I_{HF}.

Step 5: Use inverse wavelet transform to get back the fused image as $I_{F'}$.

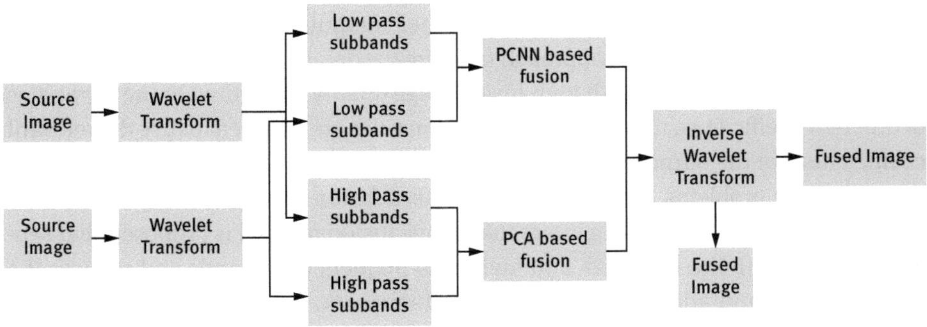

Fig. 6.9: Block diagram of proposed image fusion method [21]

6.3.3 Image fusion using PCNN and PCA in compressive sensing framework

In this proposed method, high-pass subband coefficients are fused using a PCA approach and low-pass subband coefficients are fused using PCNN. A CS framework is incorporated to reduce the processing rate and the quality of the fused image.

6.3.4 Algorithm

Step 1: Decompose source images into high-pass and low-pass subbands using wavelet transform.

Step 2: Generate measurements by projecting the high-pass and low-pass subbands into a projection matrix. Use a Gaussian measurement matrix for better efficiency.

Step 3: Recover the wavelet coefficients via a projected Landweber iteration algorithm.

Step 4: Fuse the low-pass subband using PCNN.

Step 5: Fuse the high-pass subband using a PCA-based fusion algorithm.

Step 6: Use inverse wavelet transform to get back the fused image.

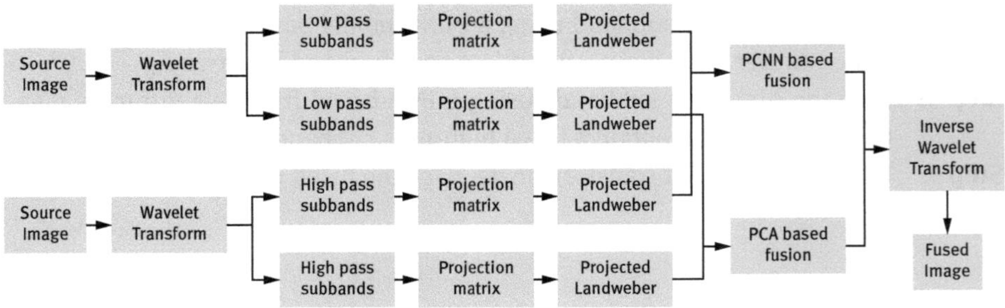

Fig. 6.10: Block diagram of proposed image fusion method

6.4 Comparative results analysis

True information and natural colour preservation are essential for any kind of image fusion method. Fusion performance is mainly evaluated using subjective and objective performance matrices. However, the results from the target data can be accurate and expensive in computational complexity and time. Hence, here the performance of different multisensor image fusion methods is evaluated as a function of their generated objective performance matrices. The following list are objective fusion measures:

1. Extract important feature points from an input image.
2. Measure the ability of the fusion process to transmit the information into an output image as accurately as possible.

From the resultant values the different performance metrics are validated from the fused images. Here all objective matrices are bounded between 0 and 1. For example, consider an objective matrix $0 \leq Q^{AB/F}(m,n) \leq 1$ where a value of 0 equals the complete loss of information, at location (m,n), as transferred from A to F or B to F, and a value of 1 corresponds to an "ideal fusion" with no loss of information. Hence for all objective performance matrices a value of 0 denotes a complete loss of information and a value of 1 means perfect natural colour fusion.

6.4.1 Objective evaluation methods

Preservation of information plays an important role in the natural appearance of fused images, and measures of fusion success must take this into account either within their total fusion efficiency or in explicit natural measures. A well-known gradient-based objective fusion evaluation framework was adopted that evaluates the preservation of input gradients in fused images to focus the investigation into metric configurations and representation spaces for comparing information between input and fused images.

In this work, the initial focus is on matrices that evaluate the performance of the fusion method that combines two input multifocus images and their resultant fused image. One of the earliest fusion-specific objective metrics was based on the idea of estimating gradient preservation. Another important body of work focuses on using mutual information and information theory measures of statistical dependence between two signals that are well suited to comparing multimodal data in general, and image quality metrics such as the structural similarity index have also been successfully applied to the evaluation of fusion, where they measure structural similarities between localized sections of each input and the fused image.

6.4.1.1 Sobel operator

The Sobel operator, typically known as a Sobel filter, is utilized in image processing and computer vision because the edges of an image are detected by its kernel function. This approximates the calculation of the gradient of an image intensity function with a discrete differentiation operator. At each point in the image, the result of the Sobel operator is based on convolving the image using a separable, small and integer valued filter in horizontal and vertical direction and is therefore relatively inexpensive in terms of computation. On the other hand, the gradient approximation produced is relatively crude, in particular for high-frequency variations in the image. The operator can use two 3×3 kernels, which are convolved with the custom image to calculate approximations of the derivatives – one for vertical changes and the other one for horizontal. Consider A as the input image, and G_x and G_y are two other images that at each factor incorporate the horizontal and vertical derivative approximations. The computations are as follows:

$$G_y = \begin{bmatrix} -1 & -2 & -1 \\ 0 & 0 & 0 \\ 1 & 2 & 1 \end{bmatrix} \cdot A \tag{6.42}$$

$$G_x = \begin{bmatrix} -1 & 0 & 1 \\ -2 & 0 & 2 \\ -1 & 0 & 1 \end{bmatrix} \cdot A \tag{6.43}$$

The gradient strength is given by

$$G = \sqrt{G_x^2 + G_y^2}\,. \tag{6.44}$$

The gradient orientation is given by

$$\theta = \tan^{-1}(G_y/G_x)\,. \tag{6.45}$$

6.4.1.2 Sigmoid function

A sigmoid function is a continuous non-linear activation function. The word sigmoid derives from the fact that the function is S-shaped. Statisticians use it in reference to a logistic function, utilizing x as an input, $f(x)$ as output with respect to t as a distinct aspect term. The sigmoid function can be given as

$$f(x) = \frac{1}{1 + e^{-tx}}\,. \tag{6.46}$$

The characteristic of the sigmoid function has a smooth continuous function whose output value ranges between −1 to 1.

6.4.1.3 Gradient preservation fusion evaluation framework

A gradient information preservation–based fusion evaluation framework, also called a $Q^{AB/F}$ metric, is associated with gradient information that is transferred from source input images to fused images. The efficiency of such a fusion algorithm depends upon the amount of gradient information transferred. In this work we have used the well-known Sobel edge operator for edge detection. Specifically, assuming two input images A and B and the resulting fused image F, a Sobel edge operator is applied to produce the strength g and orientation α ($\alpha \in |0\pi|$) information for each input and fused image pixel (m, n).

Using the foregoing parameters, relative strength and orientation change factors Δ_g and Δ_α between each fused image and input are derived, for example:

$$\Delta_g^{AF}(m,n) = \frac{g_F(m,n)}{g_A(m,n)}, \quad g_A(m,n) > g_F(m,n) \tag{6.47}$$

$$\frac{g_A(m,n)}{g_F(m,n)}, \quad g_A(m,n) \le g_F(m,n), \tag{6.48}$$

$$\Delta_\alpha^{AF}(m,n) = \frac{2}{\pi}\left| |\alpha_A(m,n) - \alpha_F(m,n)| - \frac{\pi}{2} \right| .$$

These factors are the basis of the gradient information preservation measure Q^{AF} obtained by orientation change elements and sigmoidal mapping of quality. This quantity represents the relative loss of information in the resultant fused image and its coefficients k_g, σ_g and k_α, σ_α determining the exact shape of the sigmoidal mapping:

$$Q_g^{AF}(m,n) = \Gamma_g \cdot (1 + \exp(k_g(\Delta_g^{AF}(m,n) - \sigma_g)))^{-1}, \tag{6.49}$$

$$Q_\alpha^{AF}(m,n) = \Gamma_\alpha \cdot (1 + \exp(k_\alpha(\Delta_\alpha^{AF}(m,n) - \sigma_\alpha)))^{-1}, \tag{6.50}$$

where $k_g = -15$, $\sigma_g = 0.5$, $\Gamma_g = 1.0006$ and $k_\alpha = -22$, $\sigma_\alpha = 0.8$, $\Gamma_\alpha = 1.0123$.

F represents the input A at (m, n). The gradient function is expressed as

$$Q^{AF}(m,n) = \sqrt{Q_g^{AF}(m,n) Q_\alpha^{AF}(m,n)} . \tag{6.51}$$

The total fusion performance $Q^{AB/F}$ is calculated by preserving input images Q^{AF} and Q^{BF} sum of gradient element weights where the weight elements w_A and w_B represent perceptual significance of every input image pixel. The range is $0 \le Q^{AB/F} \le 1$, where $Q^{AB/F} = 0$ implies an entire loss of input data and $Q^{AB/F} = 1$ shows a perfect combination with no loss of input data. In the simplest form, the perceptual weights w_A and w_B take the values of the corresponding gradient strength parameters g_A and g_B:

$$Q^{AB/F} = \frac{\sum_{\forall m,n} Q^{AF}(m,n)\omega_A(m,n) + Q^{BF}(m,n)\omega_B(m,n)}{\sum_{\forall m,n} \omega_A(m,n) + \omega_B(m,n)} . \tag{6.52}$$

Effectively $Q^{AB/F}$ measures the preservation of the most important input information at each location across the scene in the fused image.

6.4.1.4 Entropy

Entropy can be described as a measure of the amount of disorder in a system. It is conveyed as the spread of states that a system can adopt. A low-entropy framework possesses a small number of such states, while a high-entropy framework involves an expansive number of states.

On account of an image, these states compare to the grey-level value, which corresponds to individual pixels. For instance, in an 8-bit pixel of an image there are 256 possible states. If all states are similarly involved, as they are on account of an image that has been perfectly histogram equalized, the spread of states is a maximum, just like the entropy of an image. Then again, if the image has been thresholded, so that two exclusive states are involved, the entropy is low. In the event that the majority of pixels have the same value, the entropy of the image is zero.

With zero entropy,

$$E_{\mathrm{n}} = \sum_{i=0}^{N} p(i) \log_2 p(i) \,, \tag{6.53}$$

where N is the number of grey levels in the image and $p(i)$ is the normalized probability of occurrence of each grey level.

6.4.1.5 Standard deviation

In probability and statistics analysis, the standard deviation (SD) is represented by the symbol σ and is a measure that is used for evaluating the amount of variation or scattering of an arrangement of information qualities. An SD near 0 shows that the data points tend to be very close to the mean of the set, while a high SD indicates that the data points are spread out over a wider range of values.

When the SD value is high, it indicates a fused image at high contrast:

$$\mathrm{SD} = \sqrt{\frac{1}{MN} \sum_{i=0}^{M-1} \sum_{j=0}^{N-1} \left(I(i,j) - \bar{I} \right)^2} \,, \tag{6.54}$$

where I and $\bar{I}$ are respectively the fused and mean of the fused image whose size is $M \times N$.

6.4.1.6 Root-mean-square error

The root-mean-square deviation (RMSD) or root-mean-square error (RMSE) is used to measure the contrasts between qualities anticipated by a model or an estimator and the qualities actually observed. The RMSD represents the sample SD of the differences between observed values and predicted values. The individual differences are known as residuals, where the calculations are performed over the data sample that was used for estimation, and are called prediction errors when computed out-of-sample. The RMSD serves to sum up the magnitude of the errors in predictions for various times into a single measure of predictive power. RMSD is a decent measure of exactness, but only to compare the future errors of various models for a specific factors and not between variables, as it is scale-dependent.

RSME is a good indicator of the spectral quality of fused images.

$$\mathrm{RSME} = \sqrt{\frac{1}{MN}\sum_{i=0}^{M}\sum_{j=0}^{N}(R(i,j)-F(i,j))^2}\,, \tag{6.55}$$

where R and F are reference and fused images of size $M \times N$.

6.4.1.7 Peak signal-to-noise ratio

Peak signal-to-noise ratio, often abbreviated PSNR, is a term for finding the ratio between the maximum possible power of a signal and the power of corrupting noise that affects the fidelity of its representation. Since many signals have a wide element run, PSNR is typically communicated in terms of the logarithmic decibel scale.

Since image enhancement or improving the visual quality of a digital image can be subjective, conveying that one method provides a better quality image could vary from person to person. Therefore, it is a must to build up quantitative and observational measures to analyse the impacts of image enhancement algorithms on image quality. When the value is high, the fused and reference images are similar. A higher value indicates superior fusion.

$$\mathrm{PSNR} = 20\log_{10}\left(\frac{L^2}{\frac{I}{MN}\sum_{i=0}^{M}\sum_{j=0}^{N}(R(i,j)-F(i,j))^2}\right), \tag{6.56}$$

where R and F are reference and fused images of size $M \times N$ and L is the number of grey levels in the image.

6.4.1.8 Mutual information

A fused image should contain important information from the original image set. Clearly the idea of 'critical data' relies on the application and is hard to characterize. Mutual information (MI) is the measure of data that one datum contains about another.

MI measures the degree of dependency between two variables A and B by evaluating the distant value between the joint distribution $p_{AB}(a, b)$ and the associated probability value $p_A(a) \cdot p_B(b)$ has the complete independence by means of the relative entropy:

$$I_{AB}(a;b) = D(p_{AB}(a,b) \| p_A(a) \cdot p_B(b))\,, \tag{6.57}$$

$$I_{AB}(a;b) = \sum_{a,b} p_{AB}(a,b)\log_2 \frac{p_{AB}(a,b)}{p_A(a)p_B(b)}\,. \tag{6.58}$$

However, to calculate MI, the joint probability distribution function should be known. The joint and marginal distribution $p_{AB}(a,b)$, $p_A(a)$ and $p_B(b)$ are simply obtained by normalization of the joint and marginal histograms of both images. The joint histogram of images A and B is defined as

$$h_{AB}(a,b) = \sum_{m=1}^{M} \sum_{n=1}^{N} h_{AB}(I_A(m,n), I_B(m,n))\,; \quad a = 0, \ldots I\,; \quad b = 0, \ldots J\,, \tag{6.59}$$

where $M \times N$ is the size of the images, $I_A(m,n)$ and $I_B(m,n)$ are the intensity values of pixel (m,n) of images A and B, so that $I_A(m,n) \in [0,I]$ and $I_B(m,n) \in [0,J]$. Considering the source images A and B and the fused image F, the amount of information that F contains about A and B is estimated as

$$\mathrm{MI}_{FA}(f;a) = \sum_{a,f} p_{AF}(a,f) \log \frac{p_{AF}(a,f)}{p_A(a)p_F(f)}\,, \tag{6.60}$$

$$\mathrm{MI}_{FB}(f;b) = \sum_{b,f} p_{BF}(b,f) \log \frac{p_{BF}(b,f)}{p_B(b)p_F(f)}\,. \tag{6.61}$$

Consequently, the image fusion performance measure can be defined as

$$\mathrm{MI}_F^{AB} = \mathrm{MI}_{FA}(f;a) + \mathrm{MI}_{FB}(f;b)\,. \tag{6.62}$$

6.5 Experimental results

In this section we will show the image fusion results of several image fusion methods and their evaluation based on the aforementioned image quality metrics. We have used DWT for the decomposition of images into low-frequency and high-frequency subbands in both proposed methods. Integrated PCNN and PCA with a CS method refers to the proposed image fusion method based on PCNN and PCA in a CS framework. Existing NSCT and PCNN methods refer to image fusion methods described by H. Yin et al. [4].

From Tables 6.1, 6.2, 6.3, 6.4 and 6.5 and Figures 6.11, 6.12, 6.13, 6.14 and 6.15 we can infer that the proposed method of integrated PCNN and PCA with CS i.e., image fusion based on PCNN and PCA in a CS framework, gives better results than all other methods. We see that the spatial domain image fusion method that is PCA-based fusion does not perform well in terms of edge retention and also fails in terms of image contrast quality. Image contrast quality is increased in the proposed method since in the proposed image fusion method we directly process the pixel values of source images. Therefore the transform domain–based fusion was developed, and from the results it is clear that it is better than the spatial domain image fusion method. But the existing fusion method suffers from drawbacks like dissimilarity between fused and original images and also gives poor results in terms of PSNR values and requires high storage capacity. These artefacts were removed using the CS-based fusion method

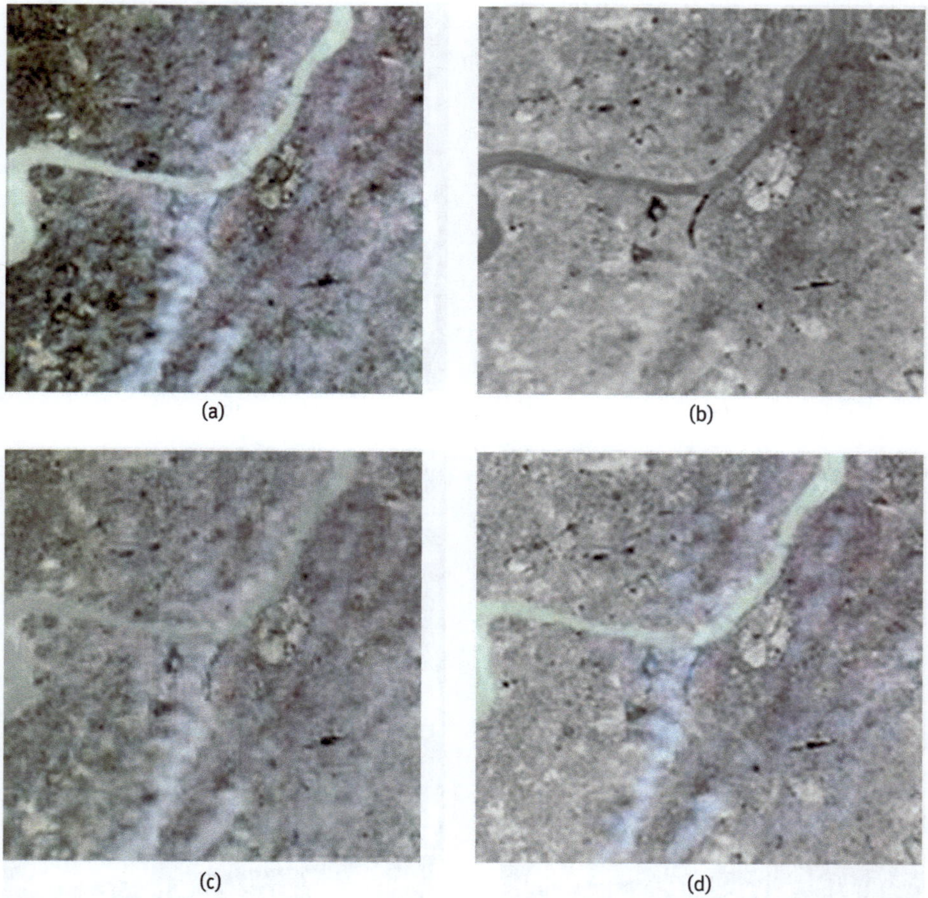

Fig. 6.11: Image fusion in CS and PCNN – aerial view of England: (a) MS image (b) PCNN (c) CS (d) Fused image

Tab. 6.1: Performance evaluation table of aerial view of England

Evaluation Metric	PCA	CS	PCNN	Integrated PCNN and PCA	Integrated PCNN and PCA with CS
$Q^{AB/F}$	0.1211	0.1454	0.1598	0.1606	0.1894
Entropy	0.3053	0.1579	0.6058	0.7016	0.7054
Standard deviation	52.7810	58.9644	82.4421	88.7211	92.4592
RMSE	3.8542	4.9387	4.9216	3.5424	3.1221
PSNR	35.6817	34.2584	42.4573	45.7992	46.7181
Mutual information	4.4310	2.3606	1.4931	4.5304	4.5564

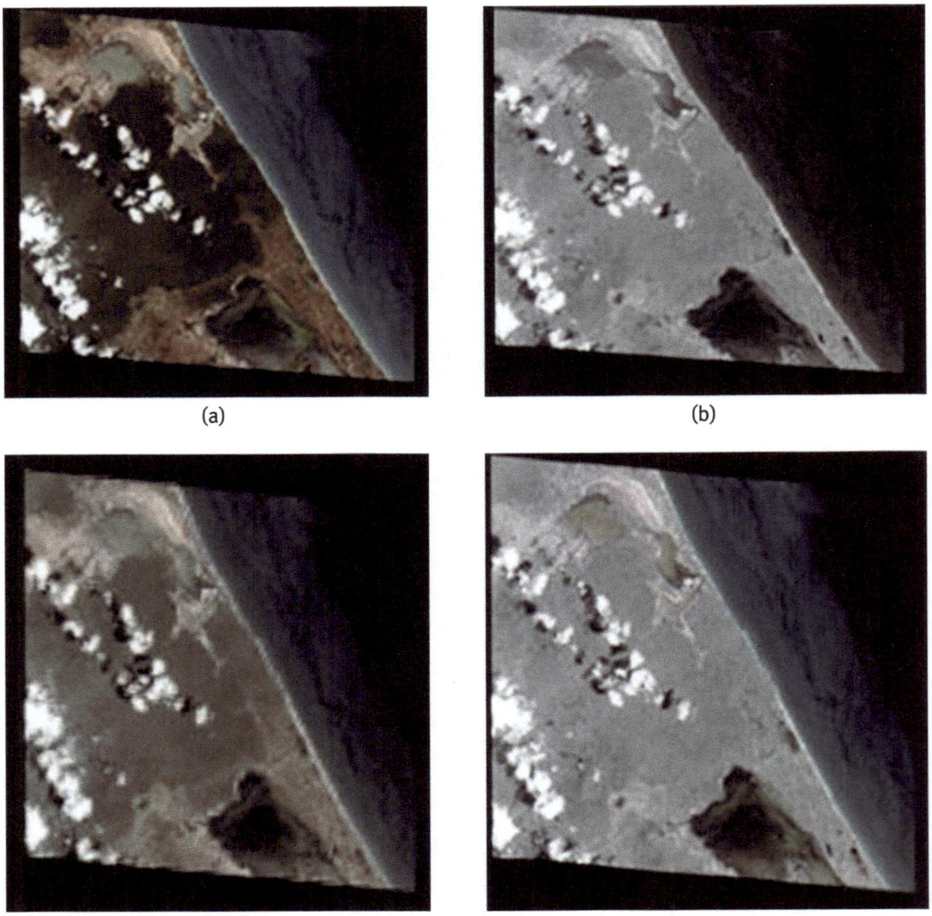

(a) (b) (c) (d)

Fig. 6.12: Image fusion in CS and PCNN – aerial view of forested land: (a) MS image (b) PCNN (c) CS (d) Fused image

Tab. 6.2: Performance evaluation table of aerial view of forested land

Evaluation Metric	PCA	CS	PCNN	Integrated PCNN and PCA	Integrated PCNN and PCA with CS
$Q^{AB/F}$	0.0911	0.0954	0.0978	0.0976	0.0970
Entropy	0.2011	0.2023	0.7046	0.7044	0.7523
Standard deviation	50.4314	55.8427	65.0933	67.9331	68.0989
RMSE	2.5378	4.1810	1.2505	1.5197	1.2231
PSNR	37.5635	35.7051	46.1893	45.9945	46.9989
Mutual information	5.8510	4.4877	5.9741	5.4707	6.7762

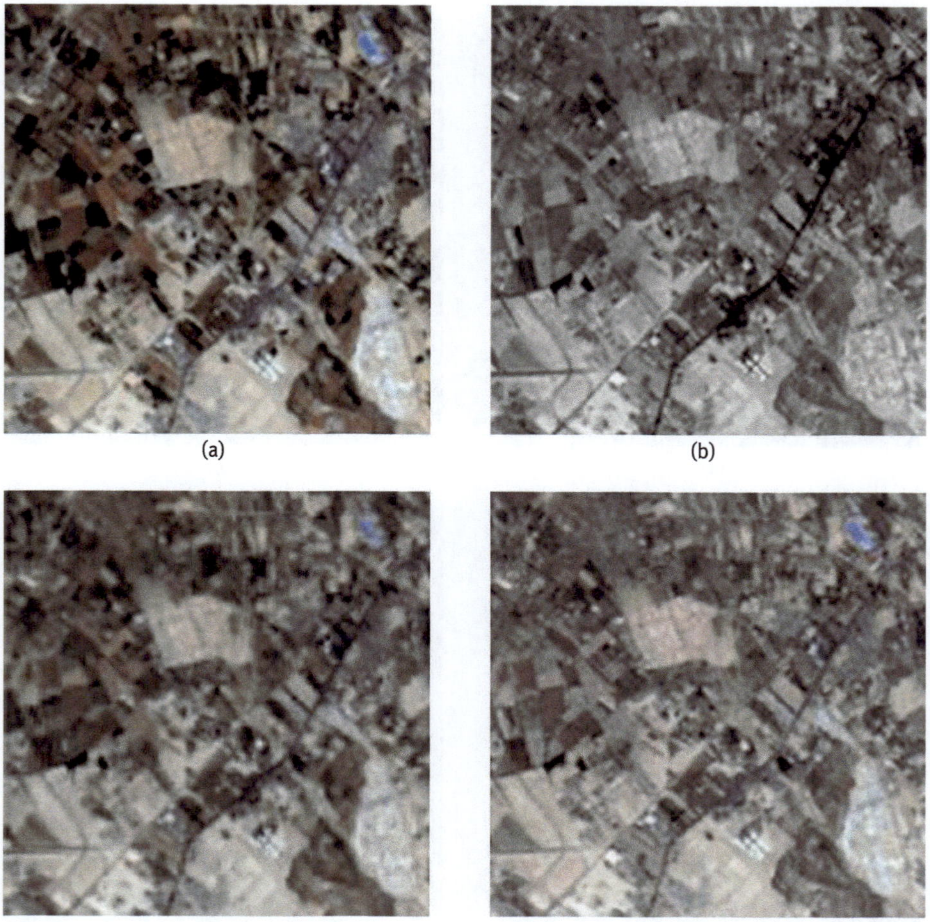

(a) (b) (c) (d)

Fig. 6.13: Image fusion in CS and PCNN – aerial view of Egypt: (a) MS image (b) PCNN (c) CS (d) Fused image

Tab. 6.3: Performance evaluation table of aerial view of Egypt

Evaluation Metric	PCA	CS	PCNN	Integrated PCNN and PCA	Integrated PCNN and PCA with CS
$Q^{AB/F}$	1.3962	1.6424	1.6121	1.7231	1.7864
Entropy	0.2601	0.2607	0.6162	0.5226	0.6224
Standard deviation	41.3671	41.4698	42.6447	42.8532	42.9672
RMSE	4.5041	3.3995	2.4313	2.2912	2.2514
PSNR	35.9141	37.5023	40.4138	40.8251	40.7627
Mutual Information	2.8234	2.6874	4.2329	5.3016	5.9065

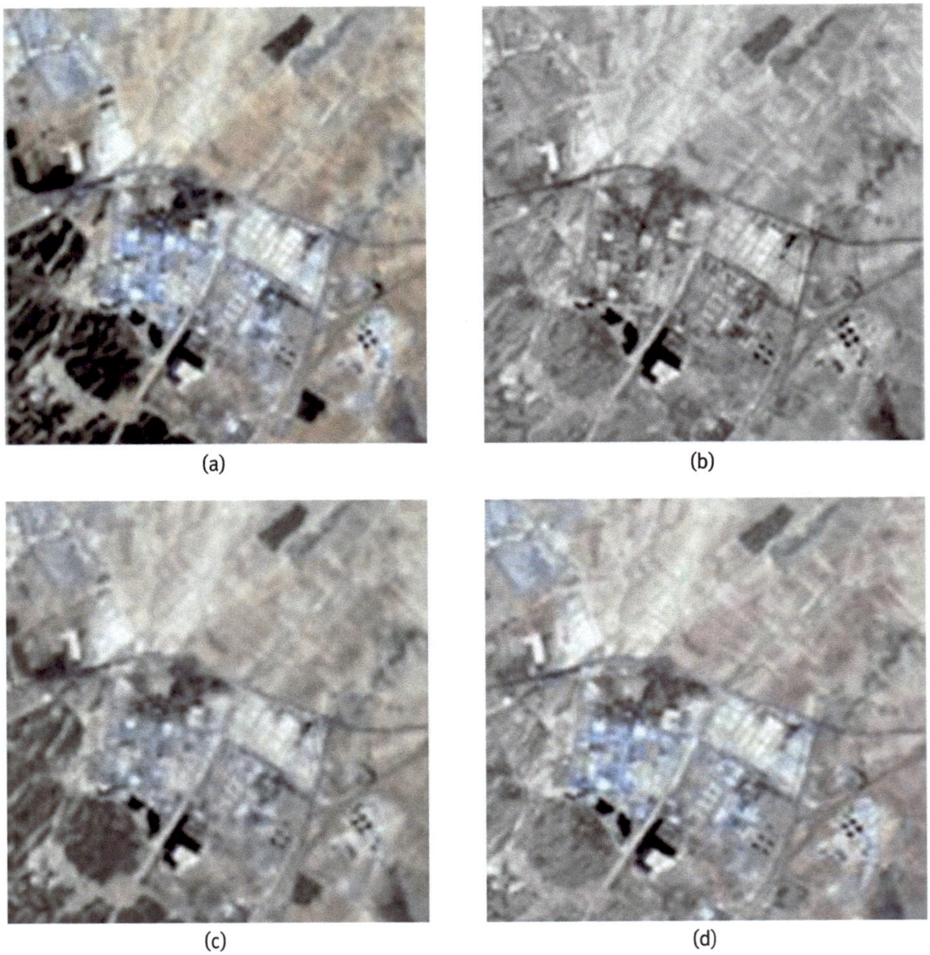

Fig. 6.14: Image fusion in CS and PCNN – aerial view of Balboa: (a) MS image (b) PCNN (c) CS (d) Fused image

Tab. 6.4: Performance evaluation table of aerial view of Balboa

Evaluation Metric	PCA	CS	PCNN	Integrated PCNN and PCA	Integrated PCNN and PCA with CS
$Q^{AB/F}$	1.6976	1.4853	1.7231	1.8244	1.7214
Entropy	0.2243	0.2702	0.6697	0.6898	0.6689
Standard deviation	42.5247	42.1781	35.1541	34.3731	31.2314
RMSE	3.4256	3.6755	2.1409	2.6591	2.6240
PSNR	36.4552	36.8244	41.5186	42.7431	42.3845
Mutual information	2.6857	2.7667	4.4716	5.3795	6.4326

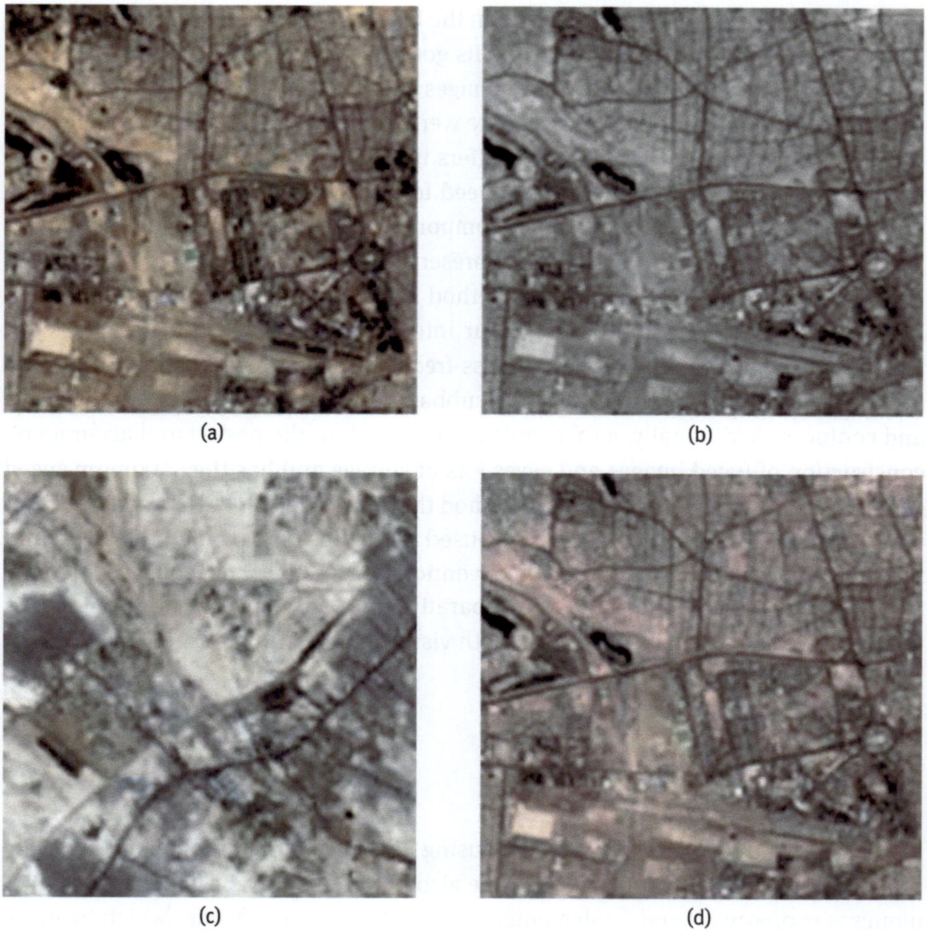

Fig. 6.15: Image fusion in CS and PCNN – aerial view of bare soil: (a) MS image (b) PCNN (c) CS (d) Fused image

Tab. 6.5: Performance evaluation table of aerial view of bare soil

Evaluation Metric	PCA	CS	PCNN	Integrated PCNN and PCA	Integrated PCNN and PCA with CS
$Q^{AB/F}$	1.2753	1.0583	1.3217	1.4284	1.3186
Entropy	0.1993	0.2656	0.6653	0.7251	0.6312
Standard deviation	40.2341	39.3149	30.8354	30.4213	29.9931
RMSE	3.3321	3.6073	1.5873	1.8546	1.7655
PSNR	35.5456	36.9873	44.1174	45.6546	44.3448
Mutual information	2.5389	2.7079	3.8677	4.5633	4.9732

where information content is conserved in the fused image. It is also computationally efficient and less expensive. But beyond its good performance, this method also suffers from artefacts like noise and fused images with poor fidelity. The CS method fails in terms of mutual information since there were no proper fusion rules for high-pass-frequency components. Hence it also suffers in preserving the gradient information in fused images. Therefore there was a need for separate image fusion rules for fusing low-pass- and high-pass-frequency components, so that the different information present in both subbands is separately preserved in the fused images. PCNN-based image fusion turns out to be the best method for fusing low-pass-frequency components since it takes account of neighbour information. PCA-based fusion turns out to be the best method for fusing high-pass-frequency components since it takes into account the variance information of the subband that effectively preserves the edges and contours. Additionally, a CS algorithm was used for the perfect and accurate reconstruction of fused images and saves lots of storage and has the maximum speed of operation. Therefore the proposed method that was based on PCNN and PCA incorporated into a CS framework can also be used for the fusion of coloured images, and its performance is better than that of conventional methods. The results were related to those of existing methods and a comparative analysis showed that the proposed method gave better results in terms of both visual quality and information content of the fused images.

6.6 Conclusion and future work

In this work, the concept of image fusion using PCNN and a CS sampling method were proposed. Image fusion applications were also discussed. A basic image fusion techniques was presented and implemented for multifocus image fusion, which includes a PCA-based fusion method, wavelet-based image fusion and so forth. Image fusion was done for the set of multifocus raw images. CS image fusion nowadays plays an important role. We also explained how PCNN- and CS-based image fusion methods are different from common image fusion methods. Integrated image fusion methods are proposed that are image fusion using PCNN and PCA with and without a CS framework. From the results it is clearly seen that the proposed image fusion method is better than the other image fusion methods. The proposed methods can also be used for the fusion of colour images. The framework was adopted with the maximum gradient that depends on objective fusion of input gradients in fused images. The proposed work is compared with a big data list of remotely sensed data in England, forested land, Egypt, Island, Balboa and bare soil, with their performance measures like $Q^{AB/F}$, entropy, SD, RMSE, PSNR and MI investigated. The results of all generated objective matrices are compared and analysed for all image fusion methods and existing image fusion methods. We saw that the proposed method based on CS image fusion gives the best performance among all image fusion methods. Image registration has a significant contribution to make in the enhancement of image fusion methods. In future, the focus should be on a multisensor image fusion method based on CS and PCNN. In future, the CS-based video fusion can also be done.

Acknowledgment: I express my gratitude to GOD ALMIGHTY for helping me in times of distress and need. I thank Dr. Deepak Mishra, Associate Professor in the department of avionics, Indian Institute of Space Technology, Trivandrum. Who gave opportunity to work with the remote sensing data for experimenting. I am indebted to Indian collaborators Dr. D. Narain Ponraj and Mr. X. Ajay Vasanth, Assistant Professor in KITS Coimbatore, India and Foreign collaborators Dr. Lawrence Henesey, Professor from Blekine Institute of Technology, Sweden and Dr. Chiung Ching Ho, Professor from Multimedia University, Malaysia who helped me to complete this dissertation as well as the challenging research that lies behind it and their meticulous guidance made this research successful.

Bibliography

[1] Xydeas CS and Petrovic V. Objective image fusion performance measure. *Electronics Letters*, 36(4):308–309, 2000.

[2] Wei S and Ke W. A multi-focus image fusion algorithm with dt-cwt. *International Conference on Computational Intelligence and Security*, pages 147–151, 2007.

[3] Pohl C and van Genderen J. Multisensor image fusion in remote sensing: Concepts, methods and applications. *International Journal of Remote Sensing*, 19:823–854, 03 1998.

[4] Yin H, Liu Z, Fang B, and Li Y. A novel image fusion approach based on compressive sensing. *Optics Communications*, 354(Supplement C):299–313, 2015.

[5] Needell D and Tropp JA. Cosamp: Iterative signal recovery from incomplete and inaccurate samples. *Applied and Computational Harmonic Analysis*, 26(3):301–321, 2009.

[6] Xiaobo Q. Image fusion algorithm based on spatial frequency motivated pulse coupled neural networks in nsct domain. *Acta Autom. Sinica*, 34:1508–1514, 01 2008.

[7] Tian Y and Tian X. Remote sensing image fusion based on orientation information in nonsubsampled contourlet transform domain. In *International Conference on Advanced Electronic Science and Technology, published*, pages 57–63. Atlantic Press, 2016.

[8] Chen Y and Qin Z. Pcnn-based image fusion in compressed domain. *Mathematical Problems in Engineering, Hindawi Publishing Corporation*, 2015.

[9] Pandit VR and Bhiwani RJ. Image fusion in remote sensing applications: A review. *International Journal of Computer Applications*, 120(10):22–32, June 2015.

[10] Li H, Ding W, Cao X, and Liu C. Image registration and fusion of visible and infrared integrated camera for medium-altitude unmanned aerial vehicle remote sensing. *Remote Sensing*, 9(5):441, 2017.

[11] Yang Y, Tong S, Huang S, Lin P, and Fang Y. A hybrid method for multi-focus image fusion based on fast discrete curvelet transform. *IEEE Translations and Content Mining*, 2017.

[12] Rajalingam B and Priya R. Multimodality medical image fusion based on hybrid fusion techniques. *International Journal on Engineering and Manufacturing Science*, 7(1):22–29, 2017.

[13] Biswas B, Sen BK, and Choudhuri R. Remote sensing image fusion using pcnn model parameter estimation by gamma distribution in shearlet domain. *Procedia Computer Science*, 70:304–310, 2015.
[14] Li J, Song M, and Peng Y. Infrared and visible image fusion based on robust principal component analysis and compressed sensing. *Infrared Physics & Technology*, 89:129–139, 2018.
[15] Selesnick IW, Baraniuk RG, and Kingsbury NC. The dual-tree complex wavelet transform. *IEEE Signal Processing Magazine*, 22(6):123–151, 2005.
[16] Eckhorn R, Reitbock HJ, Arndt M, and Dicke P. A neural network for feature linking via synchronous activity: Results from cat visual cortex and from simulations. In Cotterill RMJ, editor, *Models of Brain Function*. Cambridge University Press, 1989.
[17] Ranganath HS, Kuntimad G, and Johnson JL. Pulse coupled neural networks for image processing. In *Southeastcon '95. Visualize the Future., Proceedings., IEEE*, pages 37–43, 1995.
[18] Baig MY, Lai EMK, and Punchihewa A. Compressed sensing-based distributed image compression. *Applied Sciences*, 4(2):128–147, 2014.
[19] Unni VS. Design and implementation of multifocus image fusion algorithms in compressive sensing framework. Master's thesis, Indian Institute of Science, Trivandram, 2014.
[20] Islam SMR, Huang X, and Ou KL. Image compression based on compressive sensing using wavelet lifting scheme. *The International Journal of Multimedia and Its Applications*, 7(1):1–16, 2015.
[21] Baraniuk RG, Cevher V, Duarte MF, and Hegde C. Model-based compressive sensing. *IEEE Transactions on Information Theory*, 56(4):1982–2001, 2010.
[22] Baraniuk RG. Compressive sensing [lecture notes]. *IEEE Signal Procesing Magazine*, 24(4):118–121, 2007.

Index

https://doi.org/10.1515/9783110551433-007

Bei Fragen zur Produktsicherheit wenden Sie sich bitte an:
If you have any questions regarding product safety,
please contact:

Walter de Gruyter GmbH
Genthiner Straße 13
10785 Berlin
productsafety@degruyterbrill.com